JANE'S

W9-BVP-953

WORLD AIRCRAFT
Recognition Handbook
Derek Wood

First published in the United Kingdom in 1979
This edition published in 1989 by
Jane's Information Group Ltd, 163 Brighton Road,
Coulsdon, Surrey CR5 2NH,
United Kingdom

Second Impression 1990

Distributed in the Philippines and
the USA and its dependencies by
Jane's Information Group Inc,
1340 Braddock Place, Suite 300,
PO Box 1436, Alexandria,
Virginia 22314-1651,
USA

Jane's is a registered trade mark

ISBN 0 7106 0587 0

Computer typesetting and origination by
Method Limited, Epping, Essex, UK

Printed in the United Kingdom by
Butler and Tanner Limited, Frome, Somerset

Contents

ACKNOWLEDGEMENTS

In preparing this book I have received help from several sources: John Blake has prepared the sketch drawings, and the silhouettes have been supplied by Pilot Press and Mike Keep. I am also grateful to John Taylor and Ken Munson for the supply of several of the photographs and to Simon Elliott and Joris Janssen Lok for the model photograph. Derrick Ballington kindly produced the test sillographs, the maps and the annotated aircraft drawings.

FOREWORD TO FOURTH EDITION

Since the third edition of *Jane's World Aircraft Recognition Handbook* many new aircraft and helicopters have emerged and others have moved from prototype status to production and operational use. One entirely new aspect has emerged with low observable stealth aircraft with bizarre outlines.

At the same time a number of old faithfuls, such as the Lightning, Super Sabre, 'Fishpot' and Delta Dart, have left the scene. Others, included in the casualty list due to cancellation, are the Nimrod AEW3, the T-46A trainer, Northrop Tigershark and Learfan 2100.

The ever-rising cost of new military programmes has led to the modification and refurbishing of existing types, with consequent changes in outline and external loads. Canards have also become commonplace.

In the civil market the airline boom has led to unprecedented orders for new jet transports both twin and four engined, with winglets a common feature. New, quiet, turboprop types have also confused the recognition picture.

In this edition, the text has been revised throughout; new silhouettes and new photographs have been incorporated. New sections have been added, covering Chinese designations, an inventory of the world's combat aircraft with maps and a new class section for Stealth.

At the back of the book a 'Very New and Rather Old' section has been added. This covers new types which may go into service in the time that the reader retains this volume and older aircraft which will linger on in the same period.

INTRODUCTION

This fourth edition of *Jane's World Aircraft Recognition Handbook* comes at a time when the subject is as important as ever for military and civilian observers alike. This volume is intended as a primer for students of the subject, and is not a standard collection of aircraft types by country and manufacturer with full data. There are excellent volumes of this type such as *Jane's All the World's Aircraft*.

The aim has been to satisfy the needs of three types of recognition student:

Armed service or auxiliary personnel worldwide to whom recognition is vital for survival in war.

Those who have an interest in aviation and wish to learn how to recognise aircraft from scratch.

The knowledgeable enthusiast in service or civilian life who requires quick answers, either to check judgement or for revision. This includes civil aviation personnel.

The initial 'teach-in' section provides an introduction to the subject, while there is a glossary at the back, without which the newcomer would be lost in a world that has built up its own exclusive jargon.

The essential elements of airframes are illustrated, together with an explanation of the breakdown of the book into fixed-wing aircraft and helicopters of various distinctive types. The emphasis throughout is on shapes and not manufacturers or designations, and all groups are assembled on the basis of similarity in general configuration. After all, when an airborne object comes into view, the observer on the ground wants to refer to a shape to find out what it is. Is it jet or propeller-driven? Are its wings straight or swept? Are the engines on the wings?

The 500+ aircraft shown in the reference part of the book have been selected as most likely to be seen round the world. There is a limit to the number of aircraft that can be illustrated and those omitted include several 'one-offs', aircraft of which only a handful are flying, and ultra-lights and homebuilts.

Both military and civil machines are included, as it is impossible, even for the serviceman, just to study warplanes. Confusion can always arise if the observer has not covered all the significant types. In the early stages of the Second World War an axiom was adopted by those who were either unable or willing to learn recognition. Aircraft were divided into three categories: receding and presumed friendly; approaching and presumed hostile; and Lysanders. That the ultra-distinctive Lysander should be regarded as the only readily recognisable type is an indication of the very low standards that prevailed until recognition was really taken seriously and proper teaching methods were evolved. Faulty recognition in war has led to untold casualties which could have been avoided.

Black-and-white silhouettes have been used throughout the reference section of this book, as the detailed three-view silhouette remains the best method of encapsulating the shape and main features of any aircraft. Line drawings just do not give the same emphasis to recognition outline, although they can be useful once the student has become reasonably proficient.

All sorts of so-called mnemonic aircraft-recognition systems have been invented; one such is 'WETFUR', standing for Wings, Engine, Tail Unit, Fuselage, Undercarriage and Radiator. In practice these methods are completely useless, as recognition depends on the aircraft's total appearance and there is usually little enough time to look for individual features, let alone reciting sets of letters. Mnemonics have no place here.

Long-winded descriptions of the aircraft with a mass of complex data also leave little impression on the recognition student and only serve to confuse; they too have been omitted.

Additional information such as designation systems, code-name lists and general characteristics are very important, both for reference and to add to the general picture. They have therefore been included.

After the basic details on each type there is a list of the similar designs with which it might be confused. This has been done to stimulate interest and comparison. Some students may decide that aircraft X is much more like aircraft Y. If this comes after careful study of a number of silhouettes and photographs, then the individual's recognition knowledge will have been significantly extended.

Quick-reference symbols

At the top outside corner of each page in the following chapters you will find a quick-reference symbol. These are designed to show the main characteristics of each class of aircraft, so that the user can find his way immediately to the right chapter after catching a fleeting glimpse of, say, a delta shape in the sky. In some cases one symbol has had to stand for two classes of aircraft. It would have been difficult to suggest high or low wings with simple plan-views, just as the addition of something to indicate fixed or retractable undercarriage would have made those symbols unnecessarily complicated. One symbol has therefore been used for both high and low-wing twin-propeller aircraft, while all three classes of single propeller designs are represented by a single plan view.

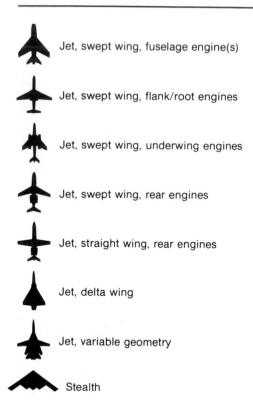

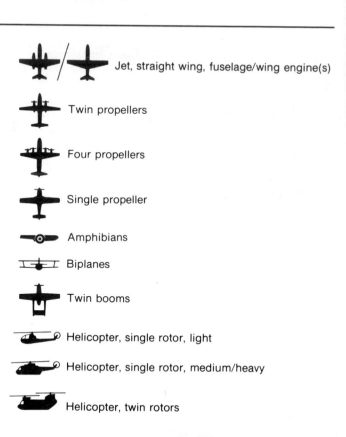

Jet, swept wing, fuselage engine(s)

Jet, swept wing, flank/root engines

Jet, swept wing, underwing engines

Jet, swept wing, rear engines

Jet, straight wing, rear engines

Jet, delta wing

Jet, variable geometry

Stealth

Jet, straight wing, fuselage/wing engine(s)

Twin propellers

Four propellers

Single propeller

Amphibians

Biplanes

Twin booms

Helicopter, single rotor, light

Helicopter, single rotor, medium/heavy

Helicopter, twin rotors

A good example of a modern twin-podded engine airliner with swept wings, the Boeing 737-400.

The very unusual Brooklands Optica observation aircraft, which looks like a jet but is in fact fitted with a piston engine driving a fan.

US Marine Corps F/A-18 Hornet with twin fins, highly swept inboard wing leading edge and multiple weapons load affecting the general outline

THE BASICS OF RECOGNITION

The initial approach to aircraft recognition is daunting, confronting the student with a mass of shapes, hundreds of odd names, curious designations, incomprehensible specifications and strange descriptive terms.

It is essential for any student to work in stages, gradually acquiring knowledge and storing it. Like people's faces and figures, every aircraft and helicopter has its own characteristics. It may have one engine and straight wings; four engines and swept wings; a fat fuselage; a triangular wing; a tailplane on top of the fin; or, in the case of a helicopter, one or two rotors.

The permutations are endless, but ultimately all powered flying machines can be sorted out into categories for detailed study. This book is divided into categories based on wing shape and engine layout in the case of fixed-wing aircraft, and size and number of rotors for helicopters.

The basic structure of an aircraft has to be studied from the outset. Almost every powered flying machine (other than helicopters) has some wing form, a fuselage and a tail structure, the last-mentioned having either a horizontal stabiliser and a fin, or a fin alone (as in the case of some delta-wing designs such as the Mirage and Concorde). Each wing, fuselage and tail structure differs in some respect from another, as do the engine installations and air intakes, be they for gas turbines or piston engines. It is the combination of these features with individual characteristics that makes up an aircraft's total shape. Knowledge of such shapes is the essence of recognition.

The same theory applies to helicopters, except that wings appear only occasionally in stub form, tail structures consist of rear rotors and some miniature tails, and the main rotor(s) dominate the shape.

The best way for the beginner to use this book is to look first at totally different types, say the MiG-21 'Fishbed' fighter and the Boeing 747 airliner, or the Jaguar ground-attack aircraft and Chinook twin-rotor helicopter. In the first case, a single-engined delta wing fighter with the engine in the fuselage just cannot resemble a giant 385+-seater with four jets slung under the wings. In the second case, the delta-wing Jaguar hardly compares with the Chinook, with its squat, boxlike fuselage and twin overhead rotors.

This process should be continued at random, selecting from the different sub-sections of the book. At the same time, the annotated drawings on pages 19–22 should be consulted to find out what the various parts of flying machines are and, in the case of flying controls, what they do. For an understanding of aviation terminology the glossary of terms and abbreviations at the end of this book can be used.

The second stage is for the reader to look at two of the aircraft in each sub-section which are similar in general layout but very different in detail characteristics. Take, for instance, the Corsair II and the Su-7 'Fitter A' in Section 1. The two wing shapes immediately strike the eye: that of the Corsair is broad in chord and moderately swept back, while the Su-7's has very sharp sweepback and slim chord. The Corsair fuselage is dumpy, with a distinctive 'pimple' above the nose intake. The SU-7, on the other hand, has a circular intake with a small pointed centrebody. Finally, the wing of the Corsair is shoulder-mounted while that of the Su-7 is positioned centre-fuselage. The Su-7 wing is straight in the head-on view while that of the Corsair has a marked anhedral.

After several excursions through the pages to see how shapes differ markedly or are superficially similar, the student should select his basic first 30 aircraft. If he is a member of a military or paramilitary force, he can choose the friendly and unfriendly types which he is most likely to encounter.

A mixed first 30 could consist of the following:

Military	Civil
Dassault Mirage	Airbus A300
MiG-29 'Fulcrum'	Boeing 737
Tupolev Tu-20 'Bear'	Boeing 747
McDonnell Douglas Phantom	McDonnell Douglas DC-8
McDonnell Douglas F-15	McDonnell Douglas DC-9
General Dynamics F-16	McDonnell Douglas DC-10
SEPECAT Jaguar	Fokker 100
Yakovlev Yak-28P 'Firebar'	Lockheed TriStar
Panavia Tornado	Tupolev Tu-154 'Careless'
Saab Viggen	Piper Comanche
Mikoyan MiG-25 'Foxbat'	Yakovlev Yak-40 'Codling'
BAe Sea Harrier	Fokker F.27 Friendship
Mikoyan MiG-21 'Fishbed'	BAe 748
Boeing B-52	Piper Cherokee
Tu-22M/Tu-26 'Backfire'	Beech Mentor

For helicopters an initial eight could be:

Military	Civil
Westland Lynx	Ka-26 'Hoodlum'
Ka-25 'Hormone'	Aérospatiale Alouette III
Boeing Chinook	Hughes 500
Sikorsky Black Hawk	MBB BO105

The student who wants to devise his own programme could well make his first selection from the types which fly from the nearest civil or military airfield, or both. After the first 30, a further 30 can be chosen, and so on. After the student has mastered his first 30 aircraft, he will be able to apply his self-acquired training to any types of his choice.

The Lockheed S-3A Viking carrier-borne anti-submarine aircraft. It has a shoulder-mounted swept wing, twin underwing turbofans and a swept fin.

11

The Nanchang Q-5, a Chinese development of the MiG-19 with flank intakes.

The only way to progress in aircraft recognition is to 'look and learn' as frequently as possible. This means not only scanning the skies but examining every photograph or drawing in newspapers and magazines and watching for aircraft on television. A scrapbook of picture cuttings and notes is an excellent way of improving the art.

A typical Soviet twin-turbine helicopter, the Mi-8 'Hip F' with single main rotor and fixed tricycle undercarriage.

Throughout, the recognition of an aircraft must involve a total impression of the shape, but with certain distinctive elements borne in mind. When any aircraft appears it must be watched right across the sky so that it can be viewed from every angle.

In the air, the head-on view of an F-15 fighter looks very different to a side or plan view, possibly leading to confusion with the Soviet MiG-25 viewed from the rear, going away, also often provide problems, as with the Mirage and Viggen. Particularly difficult are the head-on and retreating views of the transports with twin and triple rear-engine layouts, not least the Soviet Tu-134 and Tu-154 airliners. The Soviet Tu-16 and M-4 bombers are easily confused at a distance, when the difference in size cannot be estimated.

Helicopters require special study as, at a distance and in poor weather, their shapes tend to be ill-defined. In certain cases very detailed knowledge of outlines is needed to differentiate between types. Examples of similar outlines are the German MBB BO105 and the Hughes 500D helicopters.

When an unknown aircraft flies over, or an uncaptioned photograph appears in a journal, note its main features and consult the appropriate section in this book. The more 'mental images' of different types that can be stored away, the easier recognition becomes.

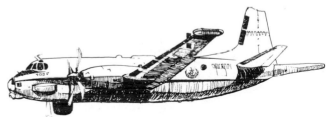

Radomes, tip tanks and tail-mounted Magnetic Anomality Detector (MAD) affect the outline of this twin-turboprop Atlantic maritime reconnaissance aircraft.

Really odd shapes. Top, the IAI 202 Arava high wing transport with twin booms, fixed undercarriage and winglets at the tips of the strut-braced wings. Below, the M-15 Belphegor agricultural biplane with twin booms, a turbofan engine and chemical hoppers between the wings.

Changing outline

Having learned the flying exterior of an aircraft, the effects of external items that may be attached to it require attention. Modern military aircraft carry a great variety of external 'stores': missiles, rocket pods, bombs, reconnaissance pods, externally mounted guns, electronic jamming pods and auxiliary fuel tanks, to mention a few. As an example of the resulting outline change, the podded fuel tanks under the wings of a Hercules transport can, at a distance, give the impression of six engines instead of four. Radomes and other bumps or bulges on an aircraft also change the shape when viewed from different angles.

Variable-geometry or swing-wing aircraft present problems, as moving the wings from straight to fully swept changes the outline continuously.

The effects of external loads on outlines. Upper left, Harrier GR5 carrying seven bombs and two Sidewinder missiles. Lower left, a Learjet with wingtip tanks, belly radome and underwing stores. Below, a Puma helicopter fitted with an Orchidée battlefield surveillance radar below the fuselage.

Finally, an aircraft preparing to land can appear completely different from its normal cruising self. For instance, a Boeing 747 cruising across the sky is not the same as a 747 on the approach to an airfield with its multi-wheel undercarriage down and massive flaps extended.

To appreciate all the changing aspects of aircraft in the air, it is essential to keep watching with the eyes or, preferably, with binoculars. When there is cloud, watch where the aircraft enters and judge where it may reappear.

Viewed from above

For the military pilot, aircraft flying below or positioned on the ground are very important. The pilot's view can be very different from that of the ground observer. It takes a sharp eye to recognise a fleeting shape parked near a hangar, in a revetment, or with its nose sticking out of a shelter. Special study is needed to separate shapes on the ground from their surroundings, and camouflage makes this even more difficult.

In the air, it is easier to recognise aircraft seen from below. From above, aircraft blend with the terrain, particularly when they are camouflaged, or have their outlines broken up by reflected light.

Contrails

At high altitude condensation of the water vapour in the exhaust gas of an aero-engine produces a 'contrail', a tell-tale stream of white vapour across the sky. On a clear day it is often possible to recognise the aircraft at the head of the trail, but usually binoculars are essential. When there is patchy cloud the point of entry of the contrail should be noted to find the direction of exit. Large multi-engined aircraft make several trails which usually blend into one. With a lot of experience it is possible to differentiate between certain types, but great care is needed.

Sound

Sound is very often the first indication that an aircraft or helicopter is approaching. Weather effects and the varied abilities of human ears can however make sound very misleading. It may appear to come from completely the wrong quarter, causing the observer to miss the approaching aircraft altogether. With fast aircraft, the subject may be well ahead of its noise. Sound is nevertheless important and certain basic facts can be gleaned from it. It is usually possible to differentiate between a piston engine and a jet; a turboprop is distinctive, while the beat of a helicopter rotor sounds different from the noise of a fixed-wing aircraft.

It is important to both watch *and* listen in order to associate sound with a particular type. A great deal of practice is needed before making snap judgements on sound alone, and even then there is a higher chance of error.

TEACHING RECOGNITION

Teaching aircraft recognition, whether in the services or a club, is an exacting task calling for method, efficiency and, above all, background knowledge of the subject. The ace spotter who can recognise anything does not necessarily make a good instructor, as he may be too far removed from the novice's problems.

An instructor must be able to make the subject live and to translate flat pictures and silhouettes into 3D images which will remain imprinted on the pupil's mind; hence the need for background knowledge. The tools of the instructor's trade are silhouettes, black-and-white or colour photographs and slides, filmstrips, the slide projector, the episcope and the overhead projector. Sometimes a film can provide excellent material, though a particular view cannot be shown again as quickly as with the other devices. A blackboard on which instructor and students can put quick sketches is extremely useful.

Models are also good tools, both to make and look at. Putting a plastic kit together will leave a lasting impression of the subject's shape and characteristics. Models may also be used in conjunction with a 'shadowgraph'. The 'flash trainer' is also an effective aid, though not always easy to obtain. The essence of the flash trainer is a shutter which flashes the image onto the screen for a brief period. With a little ingenuity it is possible to adapt an old camera shutter for use with either an episcope or a slide projector.

In approaching recognition training, it is essential to plan ahead and not swamp the pupils with an excess of information. An initial test will decide the standard of the class, while discussion will reveal the level of understanding of terms and nomenclature. A large sketch of the basic elements of fixed-wing aircraft and helicopters is a must for absolute beginners.

As mentioned earlier, contrasting types should be illustrated first to give the pupil some 'feel'; thereafter types should be taught in groups. The make-up of the groups will depend on the priorities and interests of the class: whether they are servicemen or civilians, for instance.

Each aircraft should be the subject of a lesson on its own, with a mixture of silhouettes and photographs or slides. When instruction has been given in a dozen types, there should be a full revision session and a test. As the pupils become more

Plastic models provide an excellent means of learning aircraft external characteristics – both during the assembly phase and after. Shown here is a model of the SR-71.

Computer-generated aircraft images produced in the British Aerospace Microdome trainer for Rapier missile crews (l to r): MiG-21 'Fishbed K', An-12 'Cub', Su-25 'Frogfoot', Yak-28 'Brewer', and An-12 'Cub' again.

proficient, so the tests should be varied and made more difficult.

The instructor must ensure that he does not overload the class, which will cause confusion, and also that he does not bore them. Boredom has been the enemy of good recognition training since aircraft were first used in war.

As time goes on, it will become apparent that certain types are more difficult to recognise, largely because they can be confused in outline with others of similar shape. Every effort should be made to show pictures and silhouettes of these difficult aircraft at frequent intervals until the points of difference are clearly understood. Care should however be taken to see that the pupils do not get into the bad habit of recognising the photograph rather than the aircraft itself.

The instructor can use a variety of methods to hold the students' interest. If an aviation incident, military or civil, has been headlined in the newspapers or shown on television recently, a short illustrated talk on the type or types involved should be given while the subject is fresh in the pupils' minds.

If possible, the classroom walls should be decorated with aircraft illustrations, preferably large ones, which can be changed or added to at intervals. If this is not possible, sections of hardboard covered with suitable pictures and captions should be taken along to the meetings. One such board can be used for an 'aircraft of the week' to which the class can give special attention.

'Sillographs', which form good training and test material, can be made very simply by inking in photographs with a black felt pen or a brush and black ink or paint. The proficiency test at the end of this book is presented in sillograph form. Cartoons are also of assistance, as the exaggeration stresses key points of outline. To achieve greater variety, photographs from journals can be cut into pieces showing a fin, nose, engine nacelle or part of a wing, for example.

A scrapbook can be kept handy and added to by both instructor and class. Prizes can be given for success in recognition tests. Properly organised visits to airfields and attendance at air displays will make recognition training come alive.

Finally, electronic simulators are being used to provide a new dimension in aircraft recognition. Computer-generated images of remarkable accuracy can be created. The images 'fly' correctly, with scale and distance effect and the aspect changes continuously. Weather is included – from mist to various cloud levels. On the previous page is a series of computer-generated aircraft images produced in the British Aerospace Microdome trainer for Rapier missile crews.

A typical trainer: Pilatus PC-9

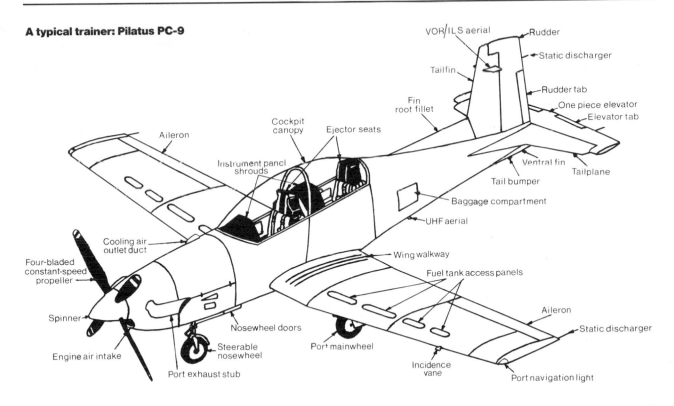

VOR/ILS aerial

Rudder

Static discharger

Tailfin

Rudder tab

One piece elevator

Elevator tab

Fin root fillet

Ventral fin

Tailplane

Tail bumper

Baggage compartment

UHF aerial

Wing walkway

Fuel tank access panels

Aileron

Static discharger

Aileron

Cockpit canopy

Ejector seats

Instrument panel shrouds

Cooling air outlet duct

Four-bladed constant-speed propeller

Spinner

Engine air intake

Port exhaust stub

Nosewheel doors

Steerable nosewheel

Port mainwheel

Incidence vane

Port navigation light

A typical airliner: Boeing 737-300

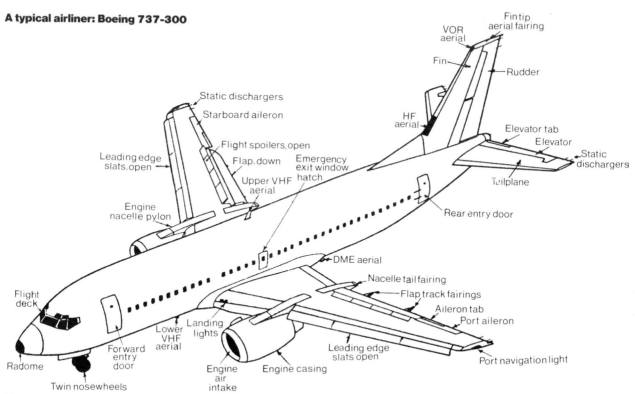

VOR aerial
Fin tip aerial fairing
Fin
Rudder
HF aerial
Static dischargers
Starboard aileron
Elevator tab
Elevator
Static dischargers
Flight spoilers, open
Leading edge slats, open
Flap, down
Emergency exit window hatch
Upper VHF aerial
Tailplane
Engine nacelle pylon
Rear entry door
DME aerial
Nacelle tail fairing
Flap track fairings
Aileron tab
Port aileron
Flight deck
Landing lights
Lower VHF aerial
Leading edge slats open
Radome
Forward entry door
Engine air intake
Engine casing
Port navigation light
Twin nosewheels

A typical helicopter: EH Industries EH 101

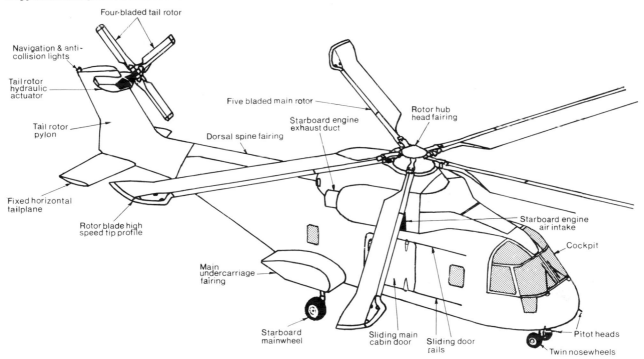

Four-bladed tail rotor

Navigation & anti-collision lights

Tail rotor hydraulic actuator

Tail rotor pylon

Five bladed main rotor

Starboard engine exhaust duct

Dorsal spine fairing

Rotor hub head fairing

Fixed horizontal tailplane

Rotor blade high speed tip profile

Starboard engine air intake

Cockpit

Main undercarriage fairing

Starboard mainwheel

Sliding main cabin door

Sliding door rails

Pitot heads

Twin nosewheels

A typical modern combat aircraft: MiG-29 'Fulcrum-A'

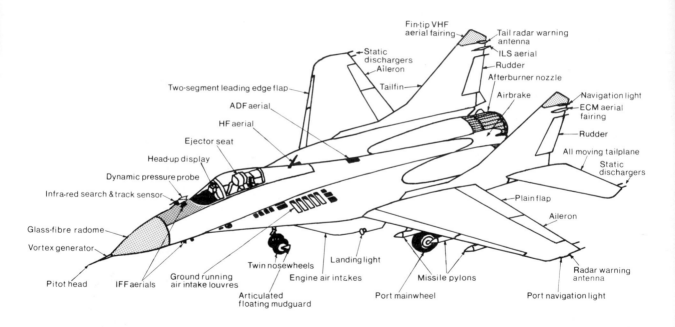

Fin-tip VHF aerial fairing

Tail radar warning antenna

ILS aerial

Rudder

Static dischargers

Aileron

Afterburner nozzle

Two-segment leading edge flap

Tailfin

Airbrake

Navigation light

ECM aerial fairing

ADF aerial

Rudder

HF aerial

All moving tailplane

Static dischargers

Ejector seat

Head-up display

Dynamic pressure probe

Infra-red search & track sensor

Plain flap

Aileron

Glass-fibre radome

Vortex generator

Landing light

Radar warning antenna

Pitot head

IFF aerials

Twin nosewheels

Engine air intakes

Missile pylons

Port navigation light

Ground running air intake louvres

Port mainwheel

Articulated floating mudguard

GAS-TURBINE ENGINES

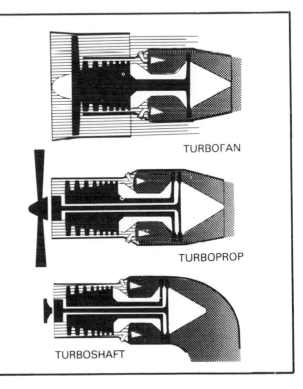

TURBOJET

TURBOFAN

TURBOPROP

TURBOSHAFT

A basic understanding of the four major classes of gas-turbine engine is an important part of aircraft-recognition training. For instance, one of the principal differences between such broadly similar types as the C-141 Starlifter and C-5A Galaxy is the fact that the former is powered by small-diameter turbofans while the latter has high-bypass-ratio turbofans, with their distinctive outsize front fans.

Gas turbines basically consist of an air compressor at the front, combustion chambers for burning the compressed air and a turbine at the back which is driven by the hot gases. In these drawings the hot-air flow is indicated by shading. The turbojet is a reaction engine which obtains its power by thrusting backwards a large weight of air. The turbofan is also a reaction engine but it differs from the turbojet in having part of the compressed air bypassed round the hot section of the engine, finally merging with the hot-air stream at the back. The turboprop has an extra turbine which drives the propeller in front. The turboshaft engine is virtually a turboprop without a propeller, the extra turbine being coupled to a shaft to drive a helicopter rotor.

GUIDED MISSILES

A wide variety of missiles is now carried as external load on military aircraft. The missiles themselves need to be recognised in order to determine the capability of the aircraft bearing them. In addition the missiles change the outline of the aircraft. The series of side views that follows shows the weapons that can form the payload of the military aircraft in this book.
(Courtesy: *Jane's Air-Launched Weapons*)

AIR-TO-SURFACE MISSILES

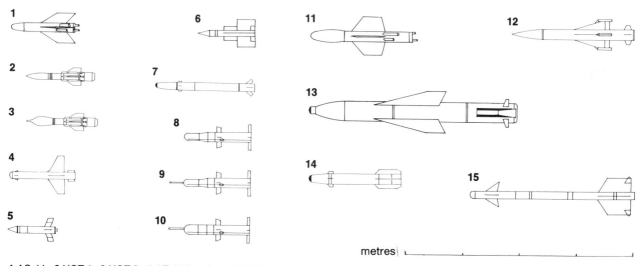

1 AS-11   **2** HOT 1   **3** HOT 2   **4** AT-2 'Swatter'   **5** AT-3 'Sagger'   **6** Mathogo   **7** AT-6 'Spiral'   **8** BGM-71A/B TOW basic
9 BGM-71C ITOW   **10** BGM-71D TOW 2   **11** AS-12   **12** AS-15   **13** AS-30L   **14** AGM-114 Hellfire   **15** AGM-122 Sidearm

AIR-TO-SURFACE MISSILES

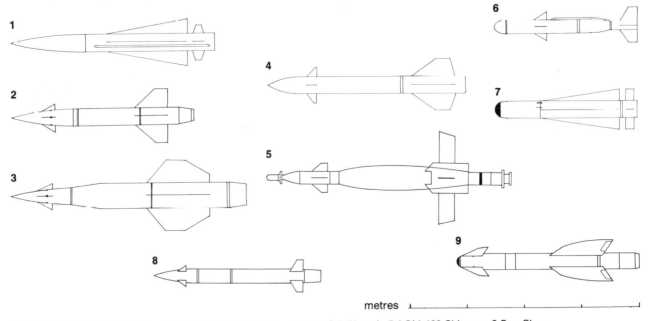

metres

1 RBO5 **2** AGM-12BE Bullpup A **3** AGM-12 C/D Bullpup B **4** AS-7 'Kerry' **5** AGM-123 Skipper **6** Sea Skua
7 AGM-65 Maverick **8** Martin Pescador **9** Penguin 3

AIR-TO-SURFACE MISSILES

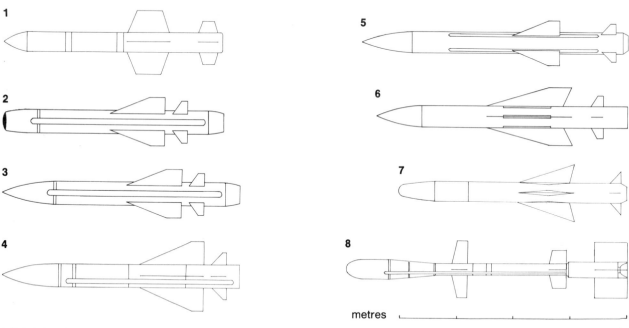

1 Gabriel 3AS 2 AJ168 Martel TV 3 AS37 Martel AR 4 Armat 5 AM-39 Exocet 6 AS-34 Kormoran 7 ASM-1 8 Marte 2A

AIR-TO-SURFACE MISSILES

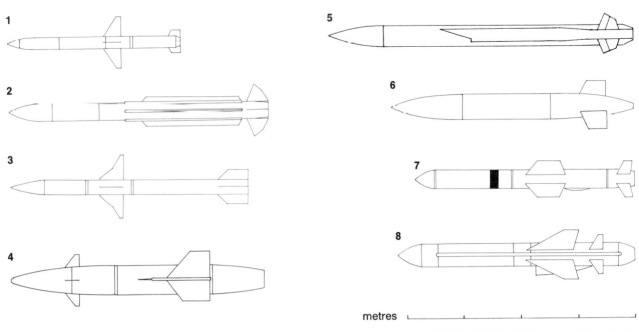

metres

1 AGM-45 Shrike   **2** AGM-78 Standard   **3** AGM-88 Harm   **4** RB04   **5** ASMP   **6** AGM-69 SRAM   **7** AGM-84 Harpoon   **8** Sea Eagle

AIR-TO-SURFACE MISSILES

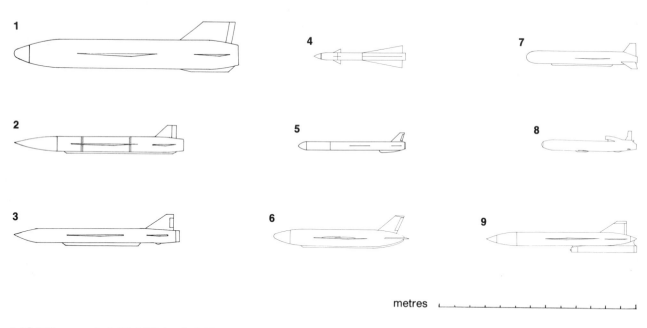

metres

1 AS-3 'Kangaroo'   **2** AS-4 'Kitchen'   **3** AS-6 'Kingfish'   **4** AS-9 'Kyle'   **5** AS-15 'Kent'   **6** AS-5 'Kelt'   **7** C-601   **8** AGM-86 ALCM
9 AS-2 'Kipper'

AIR-TO-AIR MISSILES

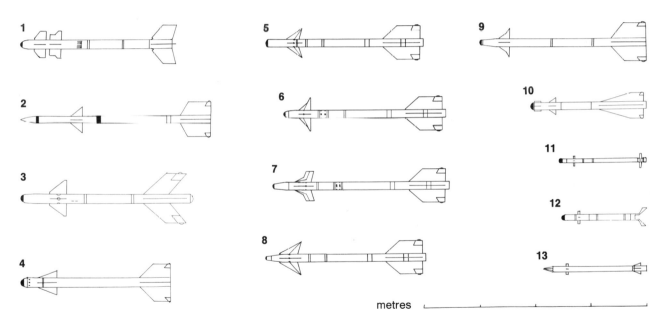

metres

1 PL-7   **2** AA-2C 'Atoll'   **3** Python   **4** Shafrir   **5** AIM-9B Sidewinder   **6** AIM-9L/M Sidewinder   **7** AIM-9P Sidewinder
8 AIM-9D Sidewinder   **9** PL-5   **10** AA-8 'Aphid'   **11** FIM-92 Stinger   **12** SA-7 'Grail'   **13** Mistral

AIR-TO-AIR MISSILES

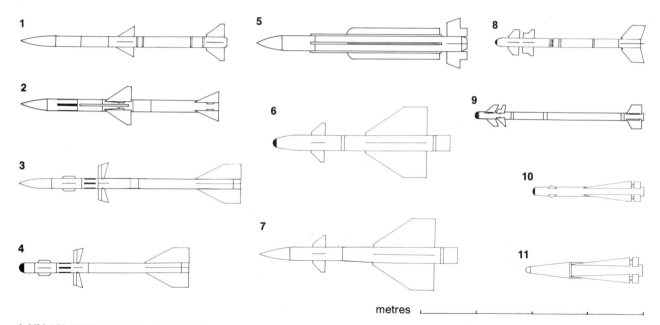

1 AIM-120 AMRAAM   **2** Sky Flash   **3** AA-10 'Alamo' radar   **4** AA-10 'Alamo' IR   **5** Super 530D   **6** AA-3 'Anab' IR
7 AA-3 'Anab' radar   **8** R-550 Magic   **9** Kukri V3   **10** AIM-4D Falcon   **11** AIM-26B Falcon

metres

AIR-TO-AIR MISSILES

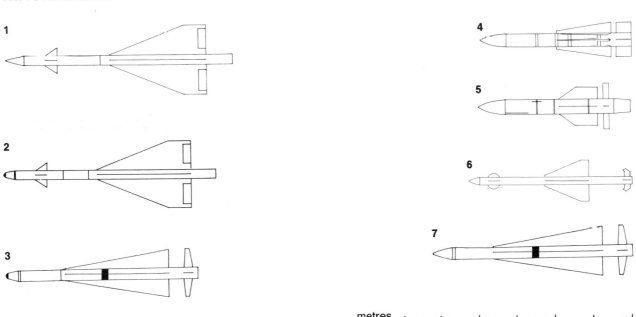

metres

1 AA-6 'Acrid' radar 2 AA-6 'Acrid' IR 3 AA-5 'Ash' IR 4 AIM-54 Phoenix 5 AA-9 'Amos' 6 AA-7 'Apex' 7 AA-5 'Ash' radar

AIRCRAFT AND GUIDED WEAPON DESIGNATIONS

The letters and numbers used to designate aircraft can be very confusing for the student. Apart from learning shapes, he has to say exactly what they are or his visual knowledge is useless. A name is always the easiest thing to remember, but in a number of cases names are not allocated or the type will appear initially with a designation before being named later. Examples of the latter are the Panavia Tornado, which began life as the MRCA (Multi-Role Combat Aircraft), and the Lockheed L-1011, which was subsequently named TriStar.

Many nations use company design numbers and/or a name, with sub-variants simply given as mark numbers. It is necessary in many cases to know the mark numbers, as the aircraft can look very different. For instance, the Boeing E-3A differs significantly from the KC-135

On the civil side there are many types which are un-named and bear only design numbers, such as the Boeing 707, 727, 737 and 747, and McDonnell Douglas DC-8, DC-9, DC-10 and MD-80. Airliners also tend to acquire suffix letters or numbers to denote different variants: 747SP, 707-320C and BAe One-Eleven 500, for example.

Five countries with special designation systems are: the United States, the Soviet Union, the United Kingdom, Canada, and China.

UNITED STATES MILITARY

The US military system appears at first to be very complicated but is in fact quite logical. It is based on the use of letters and numbers.

All United States service aircraft, guided weapons etc are assigned designations, referred to as Mission Design Series (MDS) designators. These describe the machine's basic mission, its place in the sequence designed for that function, and its place in the sequence of variants of the type. The designations are broken down as follows:

Status prefix letter This is used only when required to indicate that an aerospace vehicle is not standard because of its test, modification, experimental, or prototype design. For aircraft, the symbol appears to the immediate left of the modified mission symbol or basic mission symbol. For rockets and missiles, it is to the immediate left of the launch environment symbol or mission symbol.

Modified Mission (aircraft only) This letter is used only when needed to identify modifications to the basic mission of an aircraft and appears to the immediate left of the basic mission letter. Only one modified mission symbol is used in any one MDS.

Launch Environment (rockets and missiles only) This symbol identifies the launch environment or platform parameters. It appears to the immediate left of the mission symbol. Only one of these symbols is used in any one MDS.

Basic Mission (aircraft only) This letter identifies an aircraft's primary function or capability. It appears to the immediate left of the vehicle type letter or design number separated by a dash.

Vehicle Type (aircraft only) This letter is required only for rotary wing, vertical or short take-off/landing (VTOL/STOL) and glider aircraft and is accompanied by a basic mission or modified mission letter. It appears to the immediate left of the design number.

Mission (rockets and missiles only) This letter identifies the basic function or capability of the rocket or missile. It appears to the immediate left of the rocket- or missile-type symbol.

Vehicle Type (rockets and missiles only) This letter identifies the kind of unmanned vehicle. It appears to the immediate left of the design number separated by a dash.

Design Number This number identifies major design changes within the same mission category. Design numbers run consecutively beginning with '1' for each category. A dash separates the design number from the symbol to its immediate left.

Series This letter identifies the first production model of a particular design and later models representing major modifications that alter significantly the relationship of the vehicle to its nonexpendable system components or change its logistics support. Series symbols are consecutive beginning with 'A' and appear to the immediate right of the design number. To avoid confusion, the letters 'I' and 'O' are not used for this symbol.

Configuration Number (rockets and missiles only) This number is used only when denoting configuration changes affecting performance or tactics but not affecting non-expendable components or logistics support. It appears to the immediate right of the series symbol separated by a dash. Each military department determines its own method for assigning configuration numbers.

Block Number (aircraft only) This number identifies a production group of identically configured aircraft within a particular design series. The numbers are assigned in multiples of five (01, 05, and 10). Intermediate block numbers are reserved for field modifications and applied by the user military department.

Serial Number This number identifies a specific aerospace vehicle. Each military department determines its own method for assigning serial numbers.

Aerospace Vehicle MDS (Mission, Design, Series designators) for aircraft The following list outlines the symbols used in aircraft MDS.

Status Prefix G Permanently Grounded, **J** Special Test (temporary), **N** Special Test (permanent), **X** Experimental, **Y** Prototype, **Z** Planning.

Modified Mission A Attack, **C** Transport, **D** Director, **E** Special Electronics Installation, **F** Fighter, **H** Search and Rescue, **K** Tanker, **L** Cold Weather, **M** Multimission, **O** Observation,

P Patrol, Q Drone, R Reconnaissance, S Antisubmarine, T Trainer, U Utility, V Staff, W Weather.

Basic Mission A Attack, **B** Bomber, **C** Transport, **E** Special Electronic Installation, **F** Fighter, **O** Observation, **P** Patrol, **R** Reconnaissance, **S** Antisubmarine, **T** Trainer, **U** Utility, **X** Research.

Vehicle Type G Glider, **H** Helicopter, **V** VTOL/STOL, **Z** Lighter-than-air vehicle

Sample aircraft mission, design, series designator

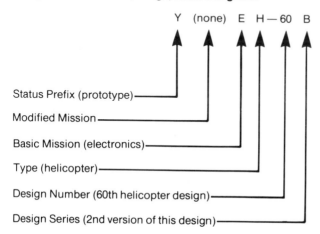

AEROSPACE VEHICLE MDS for Guided Missiles, Rockets, and Probes The following list outlines the symbols used in guided missile, rocket, and probe MDS:

Status Prefix C Captive, **D** Dummy, **J** Special Test (temporary), **M** Maintenance, **N** Special Test (permanent), **X** Experimental, **Y** Prototype, **Z** Planning.

Launch Environment A Air, **B** Multiple, **C** Coffin, **F** Individual, **G** Runway, **H** Silo Stored, **L** Silo Launched, **M** Mobile, **P** Soft Pad, **R** Ship, **U** Underwater Attack.

Sample missile mission, design, series designator

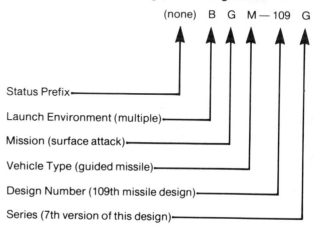

Mission D Decoy, **E** Special Electronic Installation, **G** Surface Attack, **I** Aerial Intercept, **Q** Drone, **T** Training, **U** Underwater Attack, **W** Weather.

Type M Guided Missile/Drone, **N** Probe, **R** Rocket.

While other nations have basic type names, in the United States such descriptions are known as 'popular names' when attached to a general designation. The following is the current list of aircraft and missiles so described:

Popular name	Model Designation	Service
Academe	TC-4C	Navy
Aero Commander	U-9	Navy/Army
Apache	AH-64	Army
Aquila	XMQM-105	Army
ASROC	RUR-5A	Navy
Atlas	GCM-16	AF
Aztec	U-11A	Navy
Beaver	U-6	Foreign
Black Hawk	UH-60A	Army
Bronco	OV-10	Navy/AF
Buckeye	T-2	Navy
Bulldog	AGM-83A	Navy
Bull Pup	AGM-12	Navy
Canberra	B-57	AF
Cardinal	MQM-61A	Army
Caribou	C-7	AF
Cayose	OH-6	Army
Chaparral	MIM-72	Army
Chinook	CH-47	Army
Cobra	AH-1G	Army
Cobra/TOW	AH-1Q/S	Army
Cochise	T-42A	Army
Corsair II	A-7	Navy/AF
Crusader	F-8	Navy
Dolphin	HH-65A	Coastguard
Dragonfly	A-37B	AF
Eagle	F-15	AF
Falcon	AIM-4	AF
Fighting Falcon	F-16	AF
Firebee	BQM-34	AF/Army/Navy
Fire Bolt	AQM-81	AF
Flying Classroom	T-29	Navy
Freedom Fighter	F-5A/B	AF
Galaxy	C-5A	AF
Greyhound	C-2	Navy
Guardian	HU-25A	Coastguard
Gulfstream II	VC-11A	Coastguard
HARM	AGM-88A	Navy/AF
Harpoon	AGM-84A	Navy
Harrier	AV-8	Navy
HAWK	MIM-23	Army/Navy
Hawkeye	E-2	Navy
Hellfire	AGM-114A	Army
Hercules	C-130	AF/Navy/CG
Hornet	F-18	Navy

Popular name	Model Designation	Service
Huron	C-12	AF/Army
Intruder	A-6	Navy
Iroquois	UH-1	Army
Jet Star	C-140	AF
Jolly Green Giant	HH-3E	AF
Kiowa	OH-58	Army/Navy
Lance	MGM-52	Army
Maverick	AGM-65	AF/Navy
Mentor	T-34	AF/Navy
Mescalero	T-41	Army/AF
Minuteman	LGM-30	AF
Mohawk	OV-1	Army
Neptune	P-2	Navy
Night Hawk	UH-60D	AF
Nightingale	C-9A/C	AF
Nike Hercules	MIM-14	Army
Orion	P-3	Navy
Osage	TH-55A	Army
Otter	U-1	Army
Patriot	MIM-104	Army
Peacekeeper	LGM-118A	AF
Pelican	HH-3F	CG
Pershing	MGM-31A	Army
Phantom II	F-4	Navy/AF
Phoenix	AIM-54A	Navy
Polaris	UGM-27	Navy
Provider	C-123	AF
Prowler	EA-6B	Navy

Popular name	Model Designation	Service
Raven	EF-111	AF
Redeye	FIM-43	Army
Sabre	QF-86H	Navy
Sabreliner	T-39	AF/Navy
Samaritan	C-131	AF
Sea Cobra	AH-1JT	Navy
Sea Guard	HH-52A	CG
Seahawk	SH-60B	Navy
Sea King	H-3	Navy/CG
Sea Knight	H-46	Navy
Sea Ranger	TH-57A	Navy
Sea Sparrow	RIM-7	Navy
Seasprite	H-2	Navy
Sea Stallion	CH-53A	Navy
Seminole	U-8	Army
Sentry	E-3A	AF
Shooting Star	T-33A	AF/Navy
Shrike	AGM-45	Navy/AF
Sidewinder	AIM-9	Navy/AF
Skyhawk	A-4	Navy
Skyraider	A-1	Navy
Skytrain	C-117D	Navy
Skywarrior	A-3	Navy
Sparrow	AIM-7	Navy/AF
SRAM	AGM-69	AF
Standard	RIM-66/67	Navy
Standard ARM	AGM-78	Navy/AF
Starfighter	F-104	AF

Popular name	Model Designation	Service
Starlifter	C-141	AF
Stinger	FIM-92	Army
Stratofortress	B-52	AF
Stratofreighter	KC-97L	AF
Stratolifter	C-135	AF
Stratoliner	C-137	AF
Stratotanker	KC-135	AF
Streaker	MQM-107	Army/AF
SUBROC	UUM-44A	Navy
Super Jolly	C/HH-53B/C	AF
Super Sabre	F-100	AF
Super Stallion	CH-53E	AF
Talon	T-38	AF
Talos	RIM-8	Navy
Tarhe	CH-54	Army
Tartar	RIM-24	Navy
Terrier	RIM-2	Navy
Thunderbolt II	A-10	AF
Tiger II	F-5E/F	AF
Tomahawk	BGM-109	Navy/AF
Tomcat	F-14	Navy
TOW	BGM-71	Army
Tracker	S-2	Navy
Trader	C-1A	Navy
Trident	UGM-96A	Navy
Trojan	T-28	AF/Navy
Tweet	T-37B	AF
Twin Otter	UV-18	Army/AF

Popular name	Model Designation	Service
U.S. Roland	XMIM-115	Army
Ute	U-21	Army
Viking	S-3A	Navy

SOVIET UNION MILITARY AND CIVIL

In the Soviet Union there are a number of design bureaux from which all Soviet aircraft emanate. Security is such in the Soviet Union that many new designations, particularly military, are suspect and only a certain proportion can be confirmed. For this reason the North Atlantic Treaty Organisation countries have evolved a code-name system for Soviet aircraft and this has been widely adopted elsewhere.

The known current Soviet design bureaux are:

An	Antonov
Be	Beriev
Il	Ilyushin
Ka	Kamov
Mi	Mil
MiG	Mikoyan (originally with Gurevich)
Mya	Myasishchev
Su	Sukhoi
Tu	Tupolev
Yak	Yakovlev

Code names are allotted by NATO on the basis of function:

B   Bomber
C   Cargo/transport
F   Fighter/ground attack
H   Helicopter
M   Trainer, flying boat and maritime reconnaissance.

The code names, together with Soviet designations where known for current types, are as follows:

Code name	Designation	Description
Backfire	Tu-26 (Tu-22M)	Variable-geometry bomber
Badger	Tu-16	Twinjet bomber
Beagle	Il-28	Twinjet bomber
Bear	Tu-95/Tu-142	Turboprop bomber
Bison	M-4	Four-jet bomber
Blackjack	Tu-160	V-g bomber; four-jet replacement for 'Bear'
Blinder	Tu-22	Rear-jet bomber
Brewer	Yak-28	Strike/attack
Cab	Li-2	Soviet-built DC-3
Camber	Il-86	Wide-bodied jet transport
Camel	Tu-104	Twinjet transport
Camp	An-8	Twin-turboprop transport
Candid	Il-76	Four-jet transport
Careless	Tu-154	Rear trijet transport
Cash	An-28	Turboprop version of 'Clod'
Cat	An-10	Four-turboprop transport

Code name	Designation	Description
Charger	Tu-144	Supersonic airliner
Clank	An-30	Survey version of 'Coke'
Classic	Il-62	Rear four-jet airliner
Cleat	Tu-114	Turboprop transport version of 'Bear'
Cline	An-32	Overwing turboprop version of 'Curl'
Clobber	Yak-42	Enlarged version of 'Codling'
Clod	An-14	Light piston twin
Coach	Il-12	Piston twin
Coaler	An-72	STOL replacement for 'Curl'
Cock	An-22	Four-turboprop heavy transport
Codling	Yak-40	Trijet light transport
Coke	An-24	Twin-turboprop transport
Colt	An-2	Biplane
Condor	An-124	Very large four-engined transport similar to US Galaxy
Cookpot	Tu-124	Smaller 'Camel'
Coot	Il-18	Four-turboprop airliner
Crate	Il-14	Developed 'Coach'
Creek	Yak-12	Light aircraft
Crusty	Tu-134	Rear twinjet version of 'Cookpot'
Cub	An-12	Four-turboprop transport
Cuff	Be-30	Light STOL transport
Curl	An-26	Rear-door variant of 'Coke'
Fagot	MiG-15	First swept-wing Soviet fighter in service
Farmer	MiG-19	Supersonic fighter

Code name	Designation	Description
Fencer	Su-24	V-g bomber, same class as US F-111
Fiddler	Tu-28P	Swept-wing all-weather fighter
Firebar	Yak-28P	All-weather version of 'Brewer'
Fishbed	MiG-21	Multi-variant delta fighter
Fishpot-B/C	Su-9/11	All-weather delta fighter; Su-11 lengthened
Fitter-A	Su-7B	Swept-wing ground attack fighter
Fitter-C/D	Su-17/20/22	V-g fighter
Flagon	Su-15	Supersonic delta fighter
Flanker	Su-27	Twin jet single-seater comparable to US F-15
Flipper	—	Experimental fighter
Flogger	MiG-23/27	V-g fighter/bomber; various versions
Forger	Yak-38	Supersonic VTOL carrier fighter
Foxbat	MiG-25	Supersonic interceptor
Foxhound	MiG-31	Development of 'Foxbat'
Freehand	—	Experimental VTOL aircraft
Fresco	MiG-17	Fighter; development of 'Fagot'
Frogfoot	Su-25	Twin jet ground attack aircraft
Fulcrum	MiG-29	Twin jet similar to US F-14
Halo	Mi-26	Heavy helicopter
Hare	Mi-1	Light helicopter
Harke	Mi-10	Crane variant of 'Hook'
Havoc	Mi-28	Attack helicopter, similar to US Apache
Haze	Mi-14	Naval variant of 'Hip'
Helix	Ka-27/28	Development of 'Hormone'

Code name	Designation	Description
Hen	Ka-15	Light helicopter
Hermit	Mi-34	Sport and Training helicopter
Hind	Mi-24/25	Attack helicopter
Hip	Mi-8	Military/civil helicopter
Hog	Ka-18	Development of 'Hen'
Hokum	Ka-136	New Kamov co-axial rotor combat helicopter
Hoodlum	Ka-26	General-purpose helicopter
Hook	Mi-6	Heavy military/civil helicopter
Hoplite	Mi-2	Small turbine helicopter
Hormone	Ka-25	Naval helicopter
Hound	Mi-4	Military/civil helicopter
Madcap	An-74	An-72 'Coaler' with fin radome
Madge	Be-6	Flying boat
Maestro	Yak-28U	Trainer version of 'Firebar'/ 'Brewer'
Maiden	Su-11U	Two-seater variant of 'Fishpot'
Mail	Be-12	Turboprop maritime amphibian
Mainstay	Il-76	AWACS version of 'Candid'
Magnum	Yak-30	Aerobatic jet aircraft
Mandrake	—	High-altitude reconnaissance aircraft
Mangrove	Yak-27	Reconnaissance
Mantis	Yak-32	Jet trainer
Mascot	Il-28U	Trainer variant of 'Beagle'
Max	Yak-18	Club trainer and sporting aircraft
May	Il-38	Maritime reconnaissance version of 'Coot'

Code name	Designation	Description
Maya	L-29	Delphin trainer
Midas	Il-78	Tanker version of 'Candid'
Midget	MiG-15UTI	Trainer variant of 'Fagot'
Mongol	MiG-21U	MiG-21 trainer
Moose	Yak-11	Trainer
Moss	Tu-126	Airborne early warning aircraft
Moujik	Su-7U	Trainer variant of 'Fitter A'
Mule	Po-2	Utility biplane

CANADA MILITARY

The Canadian system of military designations is similar to that of the United States. The system employs four terms, as follows:

a. **Type Designation.** This term means a two-letter prefix followed by the three-digit number assigned to each aircraft type. These three digits are the first three digits of the five or six digits that comprise the aircraft's registration or serial number. The three type designation digits commence at 100 and each aircraft type number is prefixed by 'C' for Canadian followed by the basic mission symbol (F, T, C, P or H – see subparagraph b).

b. **Basic Mission Symbol.** This term means the letter used to indicate the basic, original function of the aircraft coded as follows: F-Fighter; T-Trainer; C-Cargo/Transport; P-Patrol/Reconnaissance/Anti-Submarine Warfare; H-Helicopter.

c. **Suffix Letter.** This term means a model change or modification to the basic aircraft where such change significantly alters the aircraft's function or capability. Suffix letters, when assigned, are allotted in consecutive order commencing with the letter A. To avoid confusion, the letters I and O are not used and the letter D is reserved for dual-seat versions of single-seat aircraft. The suffix letter, when applicable, comes directly after the type designation, without space or punctuation.

d. **Popular Name.** This term means a descriptive word name (eg, Sea King, Buffalo).

Canadian designation	Canadian service name
CC-109	Cosmopolitan
CH-113	Labrador
CH-113A	Labrador
CT-114	Tutor
CC-115	Buffalo
CF-116	CF-5
CE-117	Falcon EW
CH-118	Iroquois
CP-121	Tracker
CH-124	Sea King
CC-130	Hercules
CC-132	DHC-7
CT-133	Silver Star
CT-134	Musketeer II
CH-135	Twin-Huey
CH-136	Kiowa

Canadian designation	Canadian service name
CC-137	Boeing 707
CC-138	Twin Otter
CH-139	Kiowa III
CP-140	Aurora
CC-142	Dash-8
CT-142	Dash 8M Nav Trainer
CC-144A	Challenger 600
CC-144B	Challenger 601
CE-144	Challenger EW
CH-147	Chinook
CF-188	CF-18

CHINA

The People's Republic of China (PRC) has adopted a system of designation which involves in Western terms: (1) the abbreviation of the company which produces the aircraft or helicopter; (2) the function either in full or as an abbreviation, plus the number in a design series which appears to start in most cases at −5.

The system is complicated by the fact that machines with possible export potential carry a 'westernised' function letter in place of the full or abbreviated Chinese function word.

A typical example is the NANCHANG Quiangjiji or Quiang 5 which has a westernised designation of A (A = Attack) −5.

The manufacturers' or National Aircraft Factories' abbreviations are:

Manufacturers	Abbreviation
Chengdu Aircraft Corporation	CAC
Changhe Aircraft Manufacturing Corp	CAMC
Guizhou Aviation Industry Group Company	GAIGC
Guangzhou Orlando Helicopters Ltd	GOHL
Harbin Aircraft Manufacturing Corporation	HAMC
Huabei Machinery Plant	HMP
Nanchang Aircraft Manufacturing Company	NAMC
Shenyang Aircraft Company	SAC
Shanghai Aviation Industrial Corporation	SAIC
Shaanxi Transport Aircraft Factory	STAF
Xian Aircraft Company	XAC

Function designations are as follows with, where allocated, westernised function letters:

Chinese function designation	Function	Westernised function letter
Jianjiji	Fighter aircraft	F
Jianjiji Jiaolianji or Jianjiao	Fighter trainer	FT
Suishang Hongzhaji or Suihong	Maritime bomber (patrol seaplane)	PS
Zhishengji or Zhi	Vertical take-off aircraft (helicopter)	

Chinese function designation	Function	Westernised function letter
Yunshuji or Yun	Transport aircraft	None with Westernised 'C' designation
Qiangjiji or Qiang	Attack aircraft	A
Chuji Jiaolianji or Chujiao	Basic training aircraft	PT
Hongzhaji or Hong	Bomber	B

UNITED KINGDOM MILITARY

The UK service system of aircraft designations is the reverse of that employed in the United States in that the type name is the primary reference and role letters and mark number follow.

Role letters used by the three services are:

AEW	Airborne Early Warning
AH	Army Helicopter
AL	Army Liaison
AS	Anti-submarine
B	Bomber
B(I)	Bomber/Interdictor
B(K)	Bomber (Tanker)
B(PR)	Bomber (Photo-reconnaissance)
C	Transport
CC	Transport and Communications
D	Drone or unmanned aircraft
E	Electronic Flight Checking
F	Fighter
FGA	Fighter Ground Attack
FGR	Fighter Ground Attack Reconnaissance
FR	Fighter Reconnaissance
FRS	Fighter Reconnaissance Strike
GA	Ground Attack
GR	Ground Attack Reconnaissance
HAR	Helicopter, Air Rescue
HAS	Helicopter, Anti-submarine
HC	Helicopter, Cargo
HT	Helicopter, Training
HU	Helicopter, Utility
K	Tanker
MR	Maritime Reconnaissance
R	Reconnaissance
O	Observation
PR	Photographic Reconnaissance
R	Reconnaissance
S	Strike
T	Trainer
TT	Target Towing
TX	Training Glider
W	Weather

Typical examples are Scout AH1 (Army Helicopter Mk 1), Harrier GR3 (Ground Attack Reconnaissance Mk 3) and Sea King AEW2 Airborne Early Warning Mk 2.

Aircraft and helicopter designations current in use are as follows:

Designation	Operator	Notes
A 109A	Army	Italian helicopter
Alouette		
AH2	Army	French Alouette II
Andover		
C1	MoD (PE)	Freighter
CC2	RAF	Passenger
E3	MoD (PE)	Flight checking
BAe 125		
CC1/1A	RAF/RN	BAe 125-400
CC2/2A	RAF	BAe 125-600
CC3	RAF	BAe 125-700
600	MoD (PE)	Trials
Basset		
CC1	MoD (PE)	Only one in use
Beaver		
AL1	Army	DHC-2 built in Canada
Buccaneer		
S2	RAF	Strike
S50	SAAF	S2 export to South Africa
Bulldog		
T1	RAF	Primary trainer
Canberra		
B2	RAF/RN	Still in use
PR3	MoD	One only
T4	RAF & RN	B2 trainer
B6	RAF	One only

Designation	Operator	Notes
PR7	RAF	Improved PR3
B(I)8	MoD	One only
PR9	RAF	In service
D10	MoD	Unmanned target
D14	MoD	Version of D10
B15	RAF	B6 sold abroad
E15	RAF	Flight checking
B16	MoD	Modified B15
T17	RAF	Special B2
T18	RAF/RN	Target-tug conversion
T19	RAF	T11 conversion
T22	RN	With Buccaneer nose radar
B52	Ethiopia	As B2
T54	India	As T4
B(I)56	Peru	As B6
PR57	India	As PR7
B(I)58	India	As B(I)8
B62	Argentina	As B2
T64	Argentina	As T4
B66	India	As B15/B16
PR67	India	PR7 conversion
T67	India	PR67 conversion
B(I)68	India	B(I)8 conversion
B72	Peru	B2 conversion
T74	Peru	T4 conversion
B(I)78	Peru	B(I)8 conversion
B82	Venezuela	B2 conversion
B(I)82	Venezuela	B2 conversion
PR83	Venezuela	PR3 conversion

Designation	Operator	Notes
T84	Venezuela	T4 conversion
B(I)88	Venezuela	B(I)8 conversion
Chinook		
HC1	RAF	US CH-47
Chipmunk		
T10	RAF/Army/RN	British-built trainer
Mks 21-23	Civil	
Comet 4C	MoD (PE)	3 only
Commando	Egypt	Transport adapted from Sea King
Mk 1		
Mk 2B	Egypt/Saudi Arabia/Qatar	Transport/EW
Dakota C3	MoD (PE)	One
Dominie		
T1	RAF	BAe 125 military trainer
Gazelle		
AH1	Army & RM	Gazelle B
HT2	RN	Gazelle C
HT3	RAF	Gazelle D
HCC4	RAF	Gazelle E
Grasshopper	CCF	Glider
TX1		
Harrier		
GR3	RAF	Pegasus 103 engine Also GR3A and GR3C
T4/4A	RAF	Two-seater
T4N	RN	Two-seater
GR5	RAF	AV-8B Harrier II

Designation	Operator	Notes
T6	RAF	Modified T4
GR7	RAF	Modified GR5
Mk 50	US Marines	AV-8A
Mk 52	BAe	T4, company-owned
Mk 54	US Marines	T4
Mk 55	Spanish Navy	Known as Matador/AV-8S
Mk 58	Spanish Navy	TAV-8S
T60	Indian Navy	Two-seater
Harvard		
T2B	MoD	Three only
Hawk		
T1	RAF	Trainer
T1A	RAF	With armament
T2	RAF	Trainer
Mk 50	BAe	Demonstrator
Mk 51	Finland	Trainer
Mk 52	Kenya	Trainer/ground attack
Mk 53	Indonesia	Trainer/ground attack
Mk 60	Zimbabwe	Trainer/ground attack
Mk 61	Dubai	Trainer/ground attack
Mk 63	Abu Dhabi	Trainer/ground attack
Mk 64	Kuwait	Trainer/ground attack
Mk 65	Saudi Arabia	Trainer/ground attack
Mk 66	Switzerland	Trainer/ground attack
100	BAe	Demonstrator
200	BAe	Demonstrator, single seat

Designation	Operator	Notes
Hercules		
C1	RAF	C-130K transport
CIK	RAF	Tanker
CIP	RAF	Probe
W2	RAF	Meteorological conversion
C3	RAF	Stretched fuselage
Heron		
C4	RN	Light transport
Hunter		
F6	RAF	Few
F6A	RAF	F6 to FGA9 standard
T7	RAF	Two-seat trainer
T7A	MoD (PE)	T7 modified
T8	RN	T7 for Navy
T8A	RN	C standard
T8M	RN	Sea Harrier radar trainer
FGA9	MoD (PE)	F6 with increased weapon load
GA11	RN	Naval
PR 11	RN	One only
Mk 12	MoD (PE)	Fly-by-wire experimental
F52	Peru	Export
FGA56	India	F6 for ground attack
F58	Switzerland	F6
FGA59	Iraq	FGA9
T62	Peru	One only, F4
T66	India	T7 with modified armament
T66C	Lebanon	
T69	Iraq	Two-seat version of FGA59
T70	Lebanon	T7

Designation	Operator	Notes
FGA71	Chile	FGA9
T72	Chile	T7
FGA74	Singapore	FGA9
FR74A	Singapore	PR version of FGA9
T75	Singapore	T7
FGA76	Abu Dhabi	FGA9
FR76A	Abu Dhabi	PR version of FGA9
T77	Abu Dhabi	T7
FGA78	Qatar	FGA9
T79	Qatar	T7
FGA 80	Zimbabwe	FGA9
T81	Zimbabwe	As T7
Jaguar		
GR1	RAF	Strike
T2	RAF	Two-seat trainer
Jet Provost		
T3	RAF	Trainer
T3A	RAF	Modified avionics
T4	RAF	Small number
T5	RAF	Pressurised
T5A	RAF	Modified avionics
T51—55	Various	Export versions
Jetstream		
T1	RAF	Twin trainer
T2	RAF/RN	Modified
T3	RN	Radar under fuselage-trainer
Lynx		
AH1	Army	WG13
HAS2	RN	ASW equipment

Designation	Operator	Notes
HAS3	RN	Update
Mk 4	French Navy	Uprated engines
AH5	MoD (PE)	Development
AH7	Army	
HAS8	RN	Update
AH9	Army	(wheels)
Mk 21	Brazilian Navy	
Mk 23	Argentine Navy	
Mk 25	Royal Netherlands Navy	Dutch UH-14A
Mk 27	Royal Netherlands Navy	Dutch SH-14B
Mk 28	State of Qatar	
Mk 80	Denmark	
Mk 81	Netherlands	Dutch SH-14C
Mk 86	Norway	
Mk 88	West Germany	
Mk 89	Nigerian Navy	
Mk 90	Denmark	Engine mod
Meteor		
T7	RAF/MoD	Few
NF11	MoD (PE)	One only
D16	MoD (PE)	Three only

Designation	Operator	Notes
Nimrod		
R1	RAF	Three for special EW work
MR2	RAF	Maritime reconnaissance
MR2P	RAF	With probe
Pembroke		
C1	RAF	Transport
C(PR)1	RAF	C1 photographic
Phantom		
FG1	RAF	McDonnell Douglas F-4K
FGR2	RAF	McDonnell Douglas F-4M
F-4J	RAF	Ex US Navy
Puma		
HC1	RAF	As French SA330
Scout		
AH1	Army	As Wasp
Sea Devon		
C20	RN	Naval Devon
Sea Harrier		
FRS1	RN	Naval Harrier, nose radar
FRS2	RN	Updated FRS1
Mk 51	Indian Navy	Nose radar
Sea Heron		
C2	RN	Transport
Sea King		
AEW2	RN	Airborne early warning
HAR3	RAF	Rescue
HAS5	RN	Updated
HAS6	RN	Advanced ASW
HC4	RN	Commando Mk2

Designation	Operator	Notes
Mk 6	MoD (PE)	Trials
Mk 4X	MoD (PE)	Trials
Mk 41	West Germany	
Mk 42/ 42B/42C	India	
Mk 43	Norway	
Mk 45	Pakistan Navy	
Mk 47	Egyptian Navy	
Mk 48	Belgium	
Mk 50	Australia	
Sea Prince		
T1	RAE	One only
Sea Vixen		
FAW2	MoD (PE)	For conversion
D3	MoD (PE)	Unmanned target version
Shackleton		
AEW2	RAF	Early-warning conversion of MR2
Strikemaster		
Mk 80	Saudi Arabia	
Mk 81	South Yemen	
Mk 82	Muscat&Oman	
Mk 83	Kuwait	Similar to armed Jet Provost 5
Mk 84	Singapore	
Mk 87	Kenya	
Mk 88	New Zealand	
Mk 89	Ecuador	

Designation	Operator	Notes
Tornado		
GR1	RAF	Interdiction/strike version
F2	RAF	Air-defence variant
F3	RAF	Air-defence variant
Tristar		
K1	RAF	Tanker/passenger transport
K2	RAF	ex Pan Am
Tucano		
T1	RAF	Trainer
Vampire		
T55	Several	Export trainer
VC10		
C1	RAF	Long-range transport
K2	RAF	Converted Standard VC10
K3	RAF	Converted Super VC10
K4	RAF	Update
Victor		
K2	RAF	Tanker
Wasp		
HAS1	Export	Naval Scout AH1 (wheels)
Wessex		
HC2	RAF	Single exhaust stub
HCC4	RAF	Queen's Flight
HU5	RN & RM	Commando-carrier version
HU5C	RAF	Used in Cyprus
Mk 60	Civil	
Mk 31B	RAN	HAS1 for Australia
Mk 52	Iraq	
Mk 60	Civil	Passenger

The following pages contain maps of the World, together with lists of the nation's combat aircraft. The word 'combat' is taken to include not only fighters and bombers but also ASW/MR and AEW aircraft, trainers used for attack and counter-insurgency purposes and helicopters equipped as gun ships or for battlefield close support, counter-insurgency and anti-submarine warfare. Where some countries are not referred to, it means they have no known armed combat aircraft.

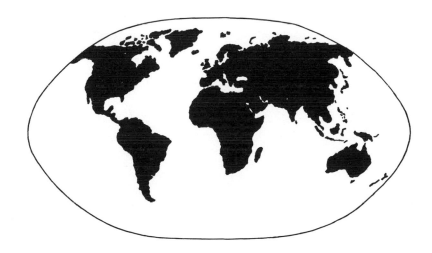

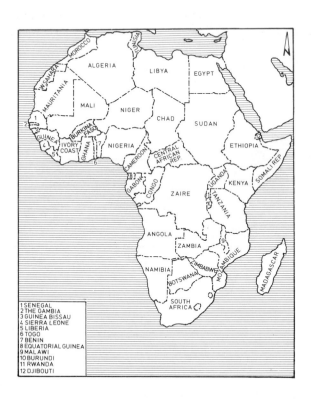

1 SENEGAL
2 THE GAMBIA
3 GUINEA BISSAU
4 SIERRA LEONE
5 LIBERIA
6 TOGO
7 BENIN
8 EQUATORIAL GUINEA
9 MALAWI
10 BURUNDI
11 RWANDA
12 DJIBOUTI

Algeria
MiG-17 'Fresco'
MiG-21 'Fishbed'
MiG-23 'Flogger'
MiG 25 'Foxbat'
Su-7/Su-20 'Fitter'
Mi-25 'Hind' (Mi-25 denotes export
    version)
Puma

Angola
MiG-17 'Fresco'
MiG-21 'Fishbed'
MiG-23 'Flogger'
Su-22 'Fitter'
Mi-8/Mi-17 'Hip'
Mi-25 'Hind'
Gazelle
SA.365M Dauphin 2

Botswana
Strikemaster
Defender

Burkina Faso
MiG-21 'Fishbed'
SF.260 Warrior

Burundi
SF.260 Warrior
SA 342 Gazelle

Cameroon
Alpha Jet
CM.170 Magister
Gazelle

Congo
MiG-17 'Fresco'
MiG-21 'Fishbed'

Djibouti
Mirage F1
(French Air Force)

Egypt
F-16
Mirage 2000
Mirage 5
Phantom
MiG-19 'Farmer'
MiG-21 'Fishbed'
Alpha Jet
Tu-16 'Badger'
Il-28 'Beagle'
E-2C Hawkeye
SA 342 Gazelle

Equatorial Guinea
MiG-17 'Fresco'

Ethiopia
MiG-17 'Fresco'
MiG-21 'Fishbed'
MiG-23 'Flogger'
Mi-8 'Hip'

Gabon
Mirage 5
Magister
Turbo-Mentor

Ghana
MB 326
MB 339
SF.260 Warrior
Alouette III

Guinea Bissau
MiG-17 'Fresco'
MiG-15 'Fagot'

Guinea Republic
MiG-17 'Fresco'
MiG-21 'Fishbed'

**Ivory Coast
(Cote D'Ivoire)**
Alpha Jet

Kenya
F-5 Tiger II
Hawk
Strikemaster
Hughes 500

Libya
MiG-21 'Fishbed'
MiG-23 'Flogger'
MiG-25 'Foxbat'
Su-22 'Fitter'
Su-24 'Fencer'
Tu-22 'Blinder'
Mirage F1
Mirage 5
Jastreb
Galeb
SF.260 Warrior
Mi-25 Hind
Gazelle

Madagascar
MiG-17 'Fresco'
MiG-21 'Fishbed'

Mali
MiG-17 'Fresco'
MiG-21 'Fishbed'

Mauritania
BN-2A Defender

Morocco
F-5
F-5 Tiger
Mirage F1
Bronco
Alpha Jet
Magister
Gazelle

Mozambique
MiG-17 'Fresco'
MiG-21 'Fishbed'

Mozambique (contd.)
Guerrier
Mi-8 'Hip'
Mi-25 'Hind'

Nigeria
MiG-21 'Fishbed'
Jaguar
Alpha Jet
Albatros
MB 339
Navy Lynx

Rwanda
Guerrier

Senegal/Gambia
CM 170 Magister
Guerrier

Somalia
Hunter
MiG-17 'Fresco'
MiG-21 'Fishbed'
Shenyang F-6 (MiG-19)
SF.260 Warrior

South Africa
Cheetah
Mirage F1
Mirage III
Buccaneer
Impala (MB 326)
Wasp
Canberra

Sudan
MiG-21 'Fishbed'
MiG-23 'Flogger'
Shenyang F-5 (MiG-17)
Shenyang F-6 (MiG-19)
F-5 Tiger II
Strikemaster
BO 105

Tanzania
Shenyang F-5 (MiG-17)
Shenyang F-6 (MiG-19)
Xian F7 (MiG-21)

Togo
Alpha Jet
EMB-326
Epsilon

Tunisia
F-5 Tiger II
MB 326
SF.260 Warrior

Uganda
SF.260 Warrior

Zaire
Mirage 5
MB 326

Zambia
MiG-21 'Fishbed'
Shenyang F-6 (MiG-19)
MB 326
Jastreb/Galeb
Warrior
AB.206

Zimbabwe
Xian F-7 (MiG-21)
Shenyang F-6 (MiG-19)
Hunter
Hawk
Warrior

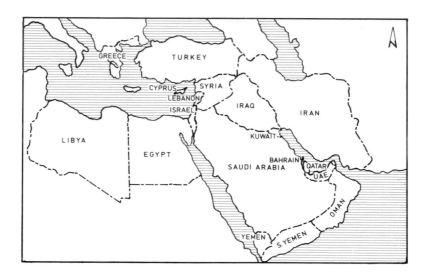

Bahrain
F-16
F-5 Tiger II

Cyprus
SA 342 Gazelle

Greece
Mirage 2000
F-16
Corsair
Starfighter
F-5
Phantom
Mirage F-1

Iran
F-14 Tomcat
Phantom
F-5 Tiger II
Xian F-7 (MiG-21)
Shenyang F-6 (MiG-19)
Orion
SH-3D Sea King
AB 212AS
Sea Cobra

Iraq
MiG-17 'Fresco'
MiG-21 'Fishbed'
MiG-23 'Flogger'
MiG-25 'Foxbat'
MiG-29 'Fulcrum'
Shenyang F-6 (MiG-19)
Su-7 'Fitter'
Su-25 'Frogfoot'
Tu-16 'Badger'
Tu-22 'Blinder'
Mirage F1
Hunter
Jet Provost
Albatros
Baghdad I (Il-76 AEW)
Super Frelon
Gazelle
Alouette
AB 212AS
BO 105
Mi-25 'Hind'

Israel
F-15
F-16
Kfir
Phantom
Skyhawk
Arava (ELINT)
Cobra
Hughes 500

Kuwait
F-18 Hornet
Mirage Fl
A-4 Skyhawk
Hawk
Super Puma
Gazelle

Lebanon
Hunter
Gazelle

Oman
Jaguar
Hunter
Strikemaster

Qatar
Mirage F1
Alpha Jet

Saudi Arabia
F-15
Tornado (IDS and ADV)
F-5 Tiger II
E-3A
Strikemaster
Dauphin

South Yemen
MiG-17 'Fresco'
MiG-21 'Fishbed'
MiG-23 'Flogger'
Su-22 'Fitter'
Il-28 'Beagle'
Mi-25 'Hind'

Syria
MiG-21 'Fishbed'
MiG-23 'Flogger'
MiG-25 'Foxbat'

Syria (contd.)
MiG-29 'Fulcrum'
Su-22 'Fresco'
Mi-25 'Hind'
Gazelle

Turkey
F-16
Starfighter
Phantom
F-5
Tracker
Cobra
AB 212 ASW

United Arab Emirates
Mirage 5
Mirage 2000
Hawk
MB 326
MB 339
Super Puma
Gazelle
Alouette

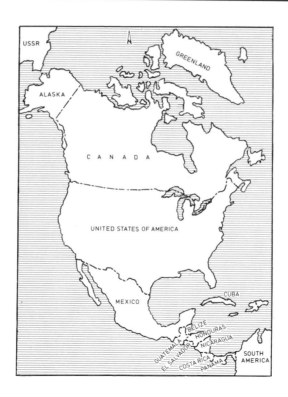

Belize
Harrier (RAF)

Canada
CF-18 Hornet
CF-5
Aurora (Orion)
CP-121 Tracker

Cuba
MiG-17 'Fresco'
MiG-19 'Farmer'
MiG-21 'Fishbed'
MiG-23 'Flogger'
Mi-8/Mi-17 'Hip'
Mi-14 'Haze'
Mi-25 'Hind'

El Salvador
Dragonfly
Magister
AC-47 Dakota (Gunship)

Guatemala
A-37B Dragonfly
Magister
Arava
Bell 212/412
Iroquois

Honduras
F-5 Tiger II
Super Mystere B.2
Sabre
Dragonfly

Mexico
F-5 Tiger II
AT-33 Shooting Star
PC-7

Nicaragua
Albatros
AT-33 Shooting Star
SF.260 Warrior
Mi-25 'Hind'

Panama
A-37B Dragonfly

USA

(USAF)
B-52 Stratofortress
B-IB
F-III/FB-III
F-4 Phantom
F-5 Tiger II
F-15 Eagle
F-16 Fighting Falcon
F-117A
Corsair II
Thunderbolt II

USA/(USAF) (contd.)
Bronco
Dragonfly
Hercules
E-3 Sentry
RC-135
TR-1

(US Army)
Apache
Cobra
Black Hawk

(USMC)
Hornet
Phantom
Skyhawk
Intruder
Prowler
Harrier
Sea Cobra

(USN)
F-18 Hornet
F-14 Tomcat
F-16 Fighting Falcon
F-5 Tiger II
Skyhawk
Corsair
Intruder
Prowler
EA-3
Hawkeye
Orion
Viking
Sea King
Seahawk
Seasprite

Argentina
Mirage III / 5 / Dagger
Super Etendard
Skyhawk
Canberra
Pucara
MB.339
Tracker
Sea King
Alouette III

Bolivia
Sabre
T-33
PC-7
Texan
Hughes 500M

Brazil
Mirage III
F-5
AMX (due for delivery)
EMB-326
Tracker
Sea King
Super Puma
Lynx

Chile
Mirage 50
F-5 Tiger II
Hunter
A-37 Dragonfly
Bandelrante (MR)
Super Puma
Dauphin

Chile (contd.)
Alouette
Jet Ranger

Colombia
Kfir
Mirage 5
A-37B Dragonfly
T-37
T-33
BO 105

Ecuador
Mirage F1
Kfir
Jaguar
Canberra
Strikemaster
A-37B Dragonfly

Paraguay
EMB-326
AT-6 Texan

Peru
Mirage 2000
Mirage 5
Su-22 'Fitter'
Canberra
A-37 Dragonfly
Tracker
Mi-25 'Hind'
Sea King
AB 212ASW
Alouette III

Uruguay
Pucara
A-37B Dragonfly
AT-33
Tracker

Venezuela
F-16
CF-5
Mirage III/5/50
Canberra
Bronco
Tracker

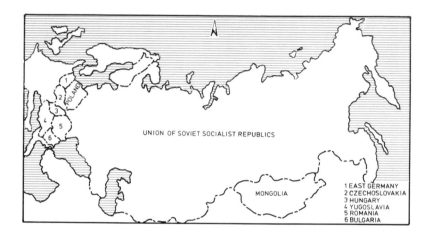

Bulgaria
MiG-17 'Fresco'
MiG-21 'Fishbed'
MiG-23 'Flogger'
Il-28 'Beagle'
Mi-4 'Hound'
Mi-8 'Hip'
Mi-14 'Haze'
Mi-24 'Hind'

Czechoslovakia
MiG-21 'Fishbed'
MiG-23 'Flogger'
Su-20/22 'Fitter'
Su-25 'Frogfoot'
Mi 24 'Hind'

East Germany
MiG-17 'Fresco'
MiG-21 'Fishbed'

East Germany (contd.)
MiG-23 'Flogger'
MiG-29 'Fulcrum'
Su-22 'Fitter'
Mi-8 'Hip'
Mi-14 'Haze'
Mi-24 'Hind'

Hungary
MiG-21 'Fishbed'
MiG-23 'Flogger'
Su-22 'Fitter'
Su-25 'Frogfoot'
Mi-24 'Hind'

Mongolia
MiG-17 'Fresco'
MiG-21 'Fishbed'

Poland
MiG-21 'Fishbed'
MiG-23 'Flogger'

Poland (contd.)
Su-7/20 'Fitter'
Il-28 'Beagle'
Iskra
Mi-14 'Haze'
Mi-24 'Hind'

Romania
MiG-17 'Fresco'
MiG-21 'Fishbed'
MiG-23 'Flogger'
Orao
Il-28 'Beagle'
Alouette III (IAR-316)
Airfox (IAR-317)

USSR
Tu-16 'Badger'
Tu-22 'Blinder'
Tu-26 'Backfire'
Tu-28 'Fiddler'

USSR (contd.)
Tu-95/142 'Bear'
Tu-160 'Blackjack'
Mya-4 'Bison'
MiG-21 'Fishbed'
MiG-23/27 'Flogger'
MiG-25 'Foxbat'
MiG-29 'Fulcrum'
MiG-31 'Foxhound'
Su-15 'Flagon'
Su-17 'Fitter'
Su-24 'Fencer'
Su-25 'Frogfoot'
Su-27 'Flanker'
Yak-28 'Firebar'
Yak-38 'Forger'
Il-76 'Mainstay'
Tu-126 'Moss'
Il-38 'May'
Be-12 'May'

USSR (cont.)
Mi-8 'Hip'
Mi-14 'Haze'
Mi-24 'Hind'
Ka-25 'Hormone'
Ka-27 'Helix'

Yugoslavia
MiG-21 'Fishbed'
MiG-29 'Fulcrum'
Orao
Jastreb
Galeb
Super Galeb
Mi-8 'Hip'
Gazelle
Ka-25 'Hormone'

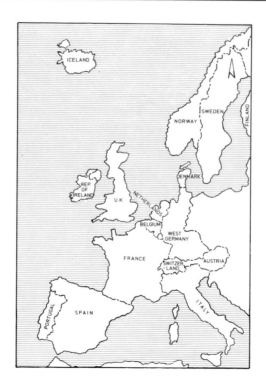

Austria
Draken
Saab 105
Belgium
F-16
Mirage 5
Denmark
F-16
Draken
AS 350L
Finland
Draken
MiG-21 'Fishbed'
France
Mirage III/5
Mirage IV
Mirage F1
Mirage 2000
Jaguar
Etendard
Super Etendard
Crusader

France (cont.)
Alize
Atlantic
Lynx
Alouette III
Gazelle
Super Puma
Italy
Tornado
Starfighter
AMX
G-91/91Y
Atlantic
Sea King
AB 212
Mangusta
Netherlands
F-16
F-5
F-27 (MR)
Orion
Lynx

Norway
F-16
F-5
Orion
Lynx
Portugal
G-91
Corsair II
Orion
Republic of Ireland
Magister
Warrior
Spain
Hornet
Phantom
Mirage F1
Mirage III
F-5
Orion
Harrier
Sea King
Seahawk

Spain (cont.)
AB 212
Hughes 500M
BO 105
Sweden
Viggen
Draken
Saab 105
Aviocar
Boeing-Vertol 107
BO 105
Switzerland
F-5
Mirage III
Hunter
United Kingdom
Tornado GR.1/F.2/F.3
Phantom
Jaguar
Harrier/Sea Harrier
Buccaneer
Nimrod

United Kingdom (cont.)
E-3 Sentry (ordered)
Shackleton
Canberra
Sea King
Lynx
West Germany
Tornado
Phantom
Starfighter
Alpha Jet
Atlantic
Sea King
Lynx
BO 105

Afghanistan
MiG 17 'Fresco'
MiG-19 'Farmer'
MiG-21 'Fishbed'
MiG-23 'Flogger'
Su-7 'Fitter'
Su-25 'Frogfoot'
Il-28 'Beagle'
Mi-8/17 'Hip'
Mi-24 'Hind'

Bangladesh
Shenyang F-6 (MiG-19)
MiG-21 'Fishbed'

Burma
AT-33
PC-7/PC-9
SF.260 Warrior

India
MiG-21 'Fishbed'
MiG-25 'Foxbat'
MiG-27 'Flogger'

India (cont.)
MiG-29 'Fulcrum'
Mirage 2000
Jaguar
Hunter
Ajeet
Marut
Sea Harrier
Canberra
Tu-142 'Bear'
Il-38 'May'
Alize

India (cont.)
Sea King
Ka-25 'Hormone'
Ka-27 'Helix'
Alouette
Mi-25 'Hind'

Pakistan
Xian F-7 (MiG-21)
Shenyang A-5 Fantan
Shenyang F-6 (MiG-19)
F-16

Pakistan (cont.)
Mirage III/5
Atlantic
Sea King
Alouette
Cobra

Sri Lanka
SF.26 Warrior
Bell 212/412
Jet Ranger

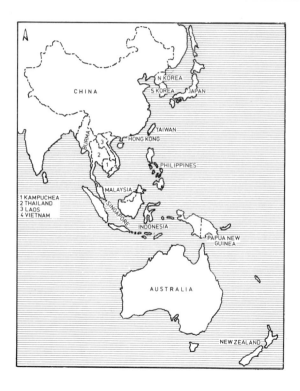

Australia
F-111
F-18 Hornet
Mirage 3
Orion
Seahawk
Sea King

Brunei
SF.260 Warrior

China (PRC)
Shenyang J-2 (MiG-15)
Shenyang J-5 (MiG-17)
Shenyang J-6 (MiG-19)
Xian J-7 (MiG-21)
Nanchang Q-5 Fantan
Shenyang J-8 Finback
Harbin H-5 (Il-28)
Xian H-6 (Tu-16)
Beriev Be-6 'Madge'
Harbin SH-5
Gazelle
Super Frelon

Indonesia
F-16
F-5 Tiger II
Skyhawk
Bronco
Hawk
Alouette
Hughes 500
Wasp

Japan
F-15 Eagle
F-4 Phantom
Mitsubishi F-1
Hawkeye
Orion
Neptune
Shin Meiwa PS-1
SH-3 Sea King
Seahawk
Cobra

Korea, North
MiG-21 'Fishbed'

Korea, North (cont.)
MiG-23 'Flogger'
MiG-29 'Fulcrum'
Shenyang F-5 (MiG-17)
Shenyang A-5 Fantan
Shenyang F-6 (MiG-19)
Su-7 'Fitter'
Su-25 'Frogfoot'
Harbin H-5 (Il-28)

Korea, South
F-16
F-4 Phantom
F-5 Tiger II
F-5
T-28 Trojan
Cobra
Hughes 500
Tracker
Lynx
Alouette

Laos
MiG-21 'Fishbed'
AC-47 Dakota (gunship)

Malaysia
Tornado (ordered)
F-5 Tiger II
Skyhawk

New Zealand
Skyhawk
Strikemaster
Orion

Singapore
F-16
F-5 Tiger II
Skyhawk
Hunter
SF.260 Warrior
Hawkeye

Taiwan
F-5 Tiger II
F-104 Starfighter
AT-3
Tracker
Hughes 500

Thailand
F-16
F-5/F-5 Tiger II
Dragonfly
Bronco
AC 47 Dakota (gunship)
AU-23 Peacemaker
Trojan
Tracker
F-27 (Maritime)
Searchmaster
Cessna/Summit 337
Bell 212

Vietnam
MiG-17 'Fresco'
MiG-19 'Farmer'
MiG-21 'Fishbed'
MiG-23 'Flogger'
Su-22 'Fitter'
F-5
A-37 Dragonfly
Il-28 'Beagle'
Be-12 'Mail'
Mi-25 'Hind'
Ka-25 'Hormone'.

CIVIL AIRCRAFT REGISTRATIONS

Registration markings are a useful aid to recognition if they can be seen. This is the list of national civil prefixes:

Prefix	Country	Prefix	Country	Prefix	Country
AP	Pakistan	EL	Liberia	MI	Marshall Islands
A2	Botswana	EP	Iran	N	United States
A5	Bhutan	ET	Ethiopia	OB	Peru
A6	United Arab Emirates	F	France	OD	Lebanon
A7	Qatar	F-O	France overseas	OE	Austria
A9C	Bahrain	G	United Kingdom	OH	Finland
A40	Oman	HA	Hungary	OK	Czechoslovakia
B	Taiwan	HB	Switzerland	OO	Belgium
BNMAU	Mongolia	HC	Ecuador	OY	Denmark
CC	Chile	HH	Haiti	P	Korea, North
CCCP	Union of Soviet Socialist Republics	HI	Dominican Republic	PH	Netherlands
		HK	Colombia	PJ	Netherlands Antilles
C/CF	Canada	HL	Korea, South	PK	Indonesia
CN	Morocco	HP	Panama	PK	West Irian
CP	Bolivia	HR	Honduras	PP, PT	Brazil
CR	Cape Verde	HS	Thailand	PZ	Surinam
CS	Portugal	HV	Vatican	P2	New Guinea
CU	Cuba	HZ	Saudi Arabia	RP	Philippines
CX	Uruguay	H4	Solomon Islands	RDPL	Laos
C2	Nauru	I	Italy	SE	Sweden
C3	Andorra	JA	Japan	SP	Poland
C5	Gambia	JY	Jordan	ST	Sudan
C6	Bahamas	J2	Djibouti	SU	Egypt
C9	Mozambique	J3	Grenada	SX	Greece
D	Germany (Fed. Rep.)	J5	Guinea-Bissau	S2	Bangladesh
DDR	Germany (Dem. Rep.)	J6	St Lucia	S7	Seychelles
DO	Fiji	J7	Dominican Republic	SN	Sao Tome
D2	Angola	J8	St Vincent	TC	Turkey
D4	Cape Verde Is	LN	Norway	TF	Iceland
D6	Comores	LQ, LV	Argentina	TG	Guatemala
EC	Spain	LX	Luxembourg	TI	Costa Rica
EI	Eire	LZ	Bulgaria	TJ	Cameroon

TL	Central African Republic	YJ	Vanuatu	6O	Somalia
TN	Congo	YK	Syria	6V	Senegal
TR	Gabon	YN	Nicaragua	6Y	Jamaica
TS	Tunisia	YR	Romania	7O	Yemen, South
TT	Chad	YS	El Salvador	7P	Lesotho
TU	Ivory Coast	YU	Yugoslavia	7Q	Malawi
TY	Benin	YV	Venezuela	7T	Algeria
TZ	Mali	Z	Zimbabwe	8P	Barbados
T2	Tuvalu	ZA	Albania	8Q	Maldives
T3	Kiribati	ZK	New Zealand	8R	Guyana
T7	San Marino	ZP	Paraguay	9G	Ghana
VH	Australia	ZS	South Africa	9H	Malta
VP-F	Falkland Islands	3A	Monaco	9J	Zambia
	(Malvinas)	3B	Mauritius	9K	Kuwait
VP-LKA/LZ	St Kitts-Nevis	3C	Guinea Equatorial	9L	Sierra Leone
VP-LMA/LUZ	Montserrat	3D	Swaziland	9M	Malaysia
VP-LVA/ZZ	British Virgin Is	3X	Guinea	9N	Nepal
VQ-T	Turks and Caicos Is	4R	Sri Lanka	9O/9T	Zaire
VR-B	Bermuda	4U	United Nations	9U	Burundi
VR-C	Cayman Islands	4W	Yemen, North	9V	Singapore
VR-G	Gibraltar	4X	Israel	9XR	Rwanda
V2	Antigua and Barbuda	4YB	Jordanian-Iraq Co-op	9Y	Trinidad and Tobago
V3	Belize		Treaty		
V8	Brunei	5A	Libya		
VR-H	Hong Kong	5B	Cyprus		
VR-U	Brunei	5H	Tanzania		
VT	India	5N	Nigeria		
XA, XB, XC	Mexico	5R	Madagascar		
XT	Burkino Faso	5T	Mauritania		
XU	Kampuchea	5U	Niger		
XV	Vietnam	5V	Togo		
XY	Burma	5W	Western Samoa		
YA	Afghanistan	5X	Uganda		
YI	Iraq	5Y	Kenya		

Jet, swept wing, fuselage engine(s)

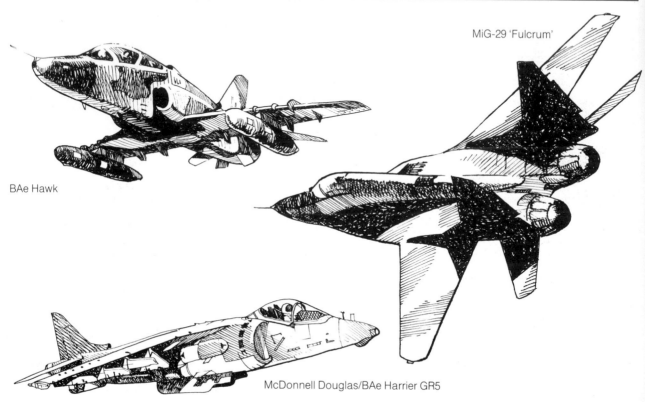

MiG-29 'Fulcrum'

BAe Hawk

McDonnell Douglas/BAe Harrier GR5

Power: 1 × Orpheus or 1 × J3 turbojet   *Span:* 10.5m   *Length:* 12.12m

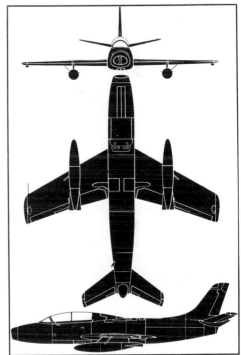

The first aircraft to be designed and built in Japan after the Second World War, the T1 Hatsutaka two-seat military trainer entered service with the Japan Air Self-Defence Force at the end of the 1950s. Bearing a superficial resemblance to the American Sabre, the T1 was initially powered by a British Orpheus engine (T1A), while later versions had the Japanese-built J3 turbojet (T1B and T1C). One machine gun is carried in the nose, and air-to-air missiles, rockets or bombs under the wings. External fuel totals 910 litres in two underwing tanks. Maximum speed of the T1A at 6,100m is 925km/h. Some 40 T1As and 22 T1Bs were built for the Japan Air Self-Defence Force. *Country of origin:* Japan. *Picture:* T1A.

 # Fiat G.91 *Confusion:* Hunter, Etendard, G.91Y

Power: 1 × Orpheus turbojet   *Span:* 8.56m   *Length:* 10.3m

Originally a Fiat design, the G.91 won a NATO light fighter-bomber design contest in the 1950s and was adopted by the Italian and Western German air forces. Internal armament is either four machine guns or two 30mm cannon. Maximum speed is 1,086km/h. The two-seat advanced trainer version is the G.91T (*upper side-view silhouette*). The G.91 is also used by Portugal. *Country of origin:* Italy. *Picture:* G.91R.

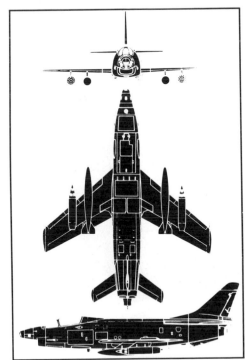

A fighter-bomber and reconnaissance aircraft, the G.91Y was a direct development of the G.91, with twin J85 turbojets replacing the single Orpheus. After a first flight in 1966, a total of 55 G.91Ys were built for the Italian Air Force. Armament comprises two fuselage-mounted 30mm cannon and four underwing pylons for rockets, bombs or guided missiles. Maximum speed at sea level is 1,110km/h. Cameras are mounted in the nose. *Country of origin:* Italy.

Mikoyan-Gurevich MiG-15UTI 'Midget'

Confusion: 'Fresco', 'Farmer'

Power: 1 × RD-45 turbojet   *Span:* 10.08m   *Length:* 10.04m

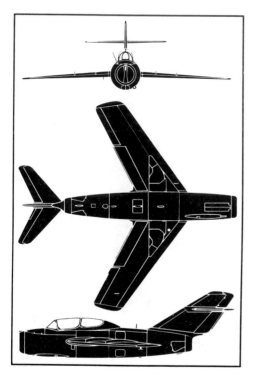

The Soviet MiG-15 swept-wing fighter, NATO code-name 'Fagot', caused a world sensation when it first appeared over the Korean War battlefields in 1950. Very large numbers of MiG-15 'Fagots' were built for use by many air forces. The MiG-15UTI two-seat advanced trainer, code-named 'Midget', was supplied throughout the Warsaw Pact and to 17 other countries. Like the 'Fagot', the 'Midget' is powered by a pirated development of the Rolls-Royce Nene. The trainer carries a cannon or a machine gun in the fuselage. *Country of origin:* USSR.

Mikoyan-Gurevich MiG-17 'Fresco'

Power: 1 × VK-1F reheated turbojet   *Span:* 9.63m   *Length:* 11.09m

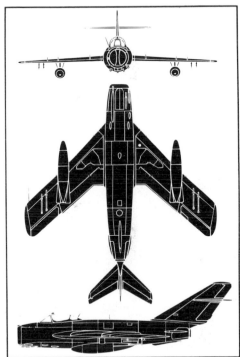

Though outwardly the MiG-17 'Fresco' bears a close similarity to the MiG-15 'Fagot', the wing and tailplane were in fact redesigned and the rear fuselage was lengthened. Later versions ('Fresco C' and 'D', 'E' and 'F') incorporated an afterburner. 'Frescos' are widely used in the fighter-bomber role by air forces round the world. Armament consists of underwing bombs and rocket pods plus three 23mm fuselage cannon on some variants and Alkali radar-homing missiles on others. 'Fresco D', 'E' and 'F' carry centrebody nose radars in addition to radar gunsights in the upper engine-intake lip. *Country of origin:* USSR. *Picture and silhouette:* 'Fresco F'.

Mikoyan MiG-19 'Farmer'

Confusion: 'Fresco'

Power: 2 × Mikulin or 2 × Tumansky reheated turbojets   *Span:* 9m   *Length:* 13.09m

The Soviet Union's first production supersonic fighter, the MiG-19 'Farmer' has been widely used and remains in service in China, Cuba, Egypt, Albania, Vietnam and Pakistan. Standard variants are 'Farmer B' (MiG-19PF and MiG-19PM with air-to-air missiles) and 'Farmer C' (MiG-19SF) with Tumansky engines. Rocket projectiles and bombs can be fitted on underwing pylons for ground attack. 'Farmer' has been built in China as the J-6 and some 2,000 are believed to be in service there. Normal gun armament is three 30mm cannon. *Country of origin:* USSR. *Silhouette:* 'Farmer C'. *Picture:* 'Farmer B'.

Power: 1 × Lyulka reheated turbojet   *Span:* 8.9m   *Length:* 17.37m

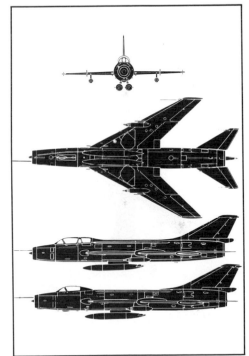

First flown in the mid-1950s, the 'Fitter A' fighter/ground attack aircraft is in service throughout the Warsaw Pact countries and in North Korea. Capable of Mach 1.6, 'Fitter' carries two 30mm cannon, rockets and bombs. External loads are carried under both fuselage and wings. Some 'Fitters' are equipped with rocket-assisted take-off gear to improve short-field performance. The two-seat trainer version, with pilot and pupil in tandem under an extended cockpit, is the Su-7U, code-named 'Moujik' (upper side view). Performance figures include a maximum level speed at 11,000m of 1,700km/h; service ceiling of 15,150m; and combat radius of 320–480km. *Country of origin:* USSR.

Nanchang Q-5 Fantan-A

Confusion: 'Farmer', 'Fitter'

Power: 2 × WP-6 reheated turbojets   *Span:* 9.7m   *Length overall:* 16.72m

In large scale service with the Chinese and Pakistan Air Forces, the Fantan-A is a twin jet attack aircraft derived from the MiG-19 'Farmer'. Compared with the 'Farmer', the Fantan has flank air intakes instead of a single nose intake. Maximum level speed is 1,190km/h and low level combat radius is 400km. There are eight external store points, of which two pairs in tandem are situated under the fuselage. Maximum bomb load is 2,000kg and two 23mm cannon are mounted in the wing roots. With an upgraded nav/attack system, designed by Aeritalia, the aircraft designation is changed to Q-5M, or A-5M in export form. *Country of origin:* China. *Picture:* A-5M.

McDonnell Douglas F-4 Phantom II

Power: 2 × J79 reheated turbojets   *Span:* 11.76m   *Length:* 18.6m

Finally phased out of production in 1979, the Phantom is probably the most numerous aircraft in Western air force inventories (over 5,000 built). Designed as an attack fighter for the US Navy, it was soon adopted by the USAF, serving in the interceptor, close support, reconnaissance (RF-4) and ECM roles. A licence-built version (F-4EJ) is in service in Japan, and Spey-engined models in the UK (F-4K/M), while certain early models in the US have been converted into pilotless target aircraft (QF-4B). A wide variety of stores can be carried, including Gatling guns, Sparrow and Sidewinder air-to-air missiles, bombs, rockets, and ECM and recce pods. Various up-date programmes for the F-4 have been carried out. *Country of origin:* USA. *Silhouette:* F-4F. *Picture:* F-4E.

SEPECAT Jaguar

Confusion: T-2/F-1, Phantom

Power: 2 × Adour reheated turbofans   *Span:* 8.49m   *Length:* 15.52m

The result of a collaborative agreement signed by the British and French governments in 1965, the Jaguar strike fighter/trainer is in large-scale service with the RAF and the Armée de l'Air. Twenty-two have been sold to Oman, 18 to Nigeria, 12 to Ecuador and a batch to India plus licence manufacture rights. Single and two-seat versions have been built. Export versions known as Jaguar International, have uprated engines and provision for the carriage of overwing air-to-air missiles in addition to twin 30mm cannon and underwing bombs, rockets and drop tanks. RAF designations for Jaguar are GR1 and T2. *Country of origin:* UK/France. *Silhouette and picture:* GR1.

Power: 2 × Adour reheated turbofans   *Span:* 7.8m   *Length:* 17.85m

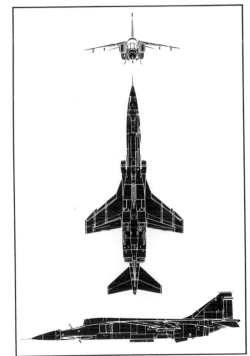

Bearing a marked resemblance to the SEPECAT Jaguar, which preceded it by some two years, the T-2 was selected for development as a two-seat supersonic trainer for the Japan Air Self-Defence Force in 1967. Ninety-four T-2 advanced and T-2A combat trainers are in service, plus 75 of a single-seat, close-support fighter variant designated F-1. F-1 carries bombs, rockets, a 20mm cannon and air-to-air missiles and has a nose radome. Powered by licence-built Rolls-Royce Adour engines similar in output to those fitted in the basic Jaguar, the T-2 is marginally faster and lighter than its European counterpart. *Country of origin:* **Japan.** *Silhouette:* F-1. *Picture:* T-2.

Dassault-Breguet Mirage F.1

Confusion: Crusader, Corsair II

Power: 1 × Atar reheated turbojet   *Span:* 8.4m   *Length:* 15m

The swept-wing Mach 2 F.1 is a single-seat multi-purpose fighter/attack aircraft, over 730 examples of which were procured by 11 countries. The F.1A and F.1E perform the ground attack role, the F.1B and F.1D are two-seat trainers, the F.1C is an interceptor and the F.1CR is for reconnaissance. The F.1C of the Armée de l'Air is equipped for air-to-air refuelling. A variety of air-to-air and air-to-ground missiles and other weapon loads can be carried in addition to two 30mm cannon. First flown in December 1966. *Country of origin:* France. *Silhouette and picture:* F.1A.

Power: 1 × turbojet + 2 × liftjets   *Span:* 7.32m   *Length:* 15.5m

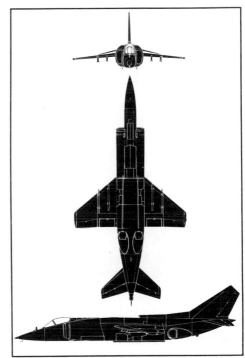

First seen aboard the aircraft carrier *Kiev* in 1976, the naval Yak-38 'Forger' was the first Soviet VTOL aircraft to enter service. It is employed in the interceptor/attack/reconnaissance roles and in its present form is just supersonic in straight-and-level flight. The powerplant layout is unusual, consisting of a single main turbojet exhausting through vectoring nozzles at the rear plus two lift engines mounted just behind the cockpit. The outer wing sections fold upwards for movement on the deck lift and hangar stowage. Armament, carried on inner wing pylons, includes guns, bombs, rockets and air-to-air missiles. A two-seat tandem trainer version has the NATO code name 'Forger B'. A successor to the Yak-38 is reported to be the Yak-41, but no details are known. *Country of origin:* USSR. *Silhouette and picture:* 'Forger A'.

British Aerospace (BAe) Harrier

Confusion: Sea Harrier, 'Forger', Etendard

Power: 1 × Pegasus vectored-thrust turbofan   *Span:* 7.70m   *Length:* (GR3): 13.91m

Of all the fixed-wing, vertical/short take-off aircraft to be conceived in the West in the late 1950s/early 1960s, the Harrier, derived from the P.1127, which first flew in 1960, is the only one to gain operational status. It is in large-scale service as a strike reconnaissance fighter with the RAF (GR3), having also served as a close support and tactical reconnaissance aircraft with the US Marine Corps (AV-8A). The Spanish Navy also operates a small number (designated Matador). Both the RAF and USMC adopted a two-seat trainer version of the aircraft with elongated fuselage (RAF T2/USMC TAV-8A). Harrier carries two 30mm Aden guns, bombs, rockets and Sidewinder air-to-air missiles. RAF Harriers have been fitted with a longer nose housing a laser rangefinder. *Country of origin:* UK. *Main silhouette and picture:* GR3; *upper side view:* T2.

Power: 1 × Pegasus vectored-thrust turbofan *Span:* 7.7m *Length:* 14.5m

Derived from the basic Harrier, the Sea Harrier FRS1 was modified for maritime operations, 57 having been ordered by the Royal Navy for operation from *Invincible* class light aircraft carriers. Principal visible differences between Sea Harrier and its land-based counterpart are a redesigned, raised cockpit and the replacement of the laser nose with a pointed radome housing the Blue Fox radar. Armament includes 30mm Aden guns, bombs, rockets, and AIM-9L air-to-air missiles. The Sea Harrier is also in carrier-borne service with the Indian Navy. Royal Navy Sea Harriers will undergo a mid-life up-date from 1991 including Blue Vixen radar and AMRAAM missiles. Span will be increased by 61cm and length by 35cm. *Country of origin:* UK. *Silhouette: SeaHarrier FRSl. Picture:* Sea Harrier FRS2.

McDonnell Douglas AV-8B/BAe Harrier GR5

Confusion: Harrier GR3, Sea Harrier, 'Forger'

Power: 1 × Pegasus vectored-thrust turbofan   *Span:* 9.24m   *Length:* 14.12m

With raised cockpit, composite-material wing with leading-edge root extensions, fuselage underside strakes and six wing weapons/fuel pylons, the AV-8B Harrier II is a second-generation V/STOL aircraft based on the Harrier formula. Jointly produced by McDonnell Douglas and BAe, the AV-8B is expected to run to 300+ examples for the US Marines plus 94 GR5s for the RAF. Compared with the existing AV-8A/Harrier GR3, the AV-8B/GR5 has a radius of action raised to 1,114km. Weapons include two 30mm cannon or a 25mm Gatling gun, bombs, cluster bombs and Maverick and Sidewinder missiles. The trainer AV-8B is designated TAV-8B. A retrofit programme is intended for the GR5 to produce a two seat night attack version – likely to be designated GR7. A single-seat night-attack version of AV-8B is already in production. Spain has procured 12 aircraft as EAV-8B. *Countries of origin:* USA/UK. *Silhouette:* AV-8B. *Picture:* GR5.

Dassault-Breguet/Dornier Alpha Jet

Power: 2 × Larzac turbofans   *Span.* 9.11m   *Length:* 12.29m

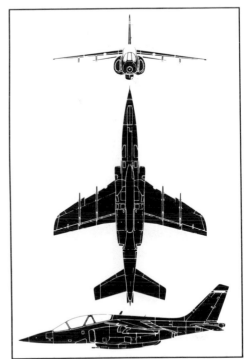

A joint venture by France and Germany, the Alpha Jet is a two-seater to be used for advanced training/light attack by the French and as the 'Close Support version' by the West Germans. First flown in 1973, the Alpha Jet has been ordered by France, West Germany and 10 other countries. Alpha Jet can carry a 30mm or 27mm gun pod under the fuselage, while the Close Support version has four wing pylons suitable for the carriage of a variety of weapons. An attack variant with improved systems and weapon load is known as Alpha Jet 2. A further improved version is called Lancier, while an improved trainer is Alpha Jet 3. One export variant has an additional dorsal spine. *Country of origin:* France/West Germany. *Silhouette:* Advanced Trainer/light attack version. *Picture:* Close Support version.

SOKO Orao/CNIAR IAR-93 Yurom

Confusion: Jaguar, Alpha Jet, Harrier

Power: 2 × Viper turbojets   *Span:* 9.62m   *Length (single seater):* 13.96m

Flown for the first time in 1974, this ground-attack fighter is a joint development by the Yugoslav and Romanian industries. The overall programme is known as 'Yurom', while the type is designated Orao 1 and 2 in Yugoslavia and IAR-93 A and B in Romania, the Orao 2 and B versions having afterburners fitted. Western suppliers provide a number of key systems. Some 400 examples of this subsonic aircraft, which has a maximum weapon load of approximately 2,000kg, are expected to be produced for the Yugoslav and Romanian air forces. Each country has also flown single prototypes of a two-seat dual-control operational trainer, and this version is also now in production. *Countries of origin:* Yugoslavia and Romania.

Aeritalia/Aermacchi/EMBRAER AM-X

Power: 1 × Spey turbofan   *Span:* 8.87m   *Length:* 13.57m

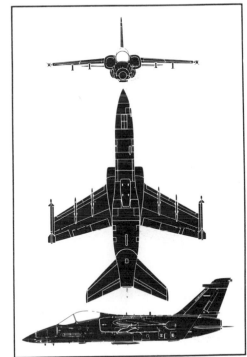

A joint Italian-Brazilian development, the AM-X is a single-seat tactical fighter-bomber which will equip the air forces of both countries, 187 for Italy and 79 for Brazil, plus 37 two-seaters. AM-X first flew in May 1984 and the first squadron was formed in Italy in June 1989. Brazilian aircraft will have two 30mm cannon while those in Italy will have a single multi-barrel 20mm gun. Other armament will include bombs, rocket projectiles and air-to-surface and air-to-air missiles. An anti-ship version is proposed, with the designation AMX-ASV. *Countries of origin:* Italy and Brazil.

Vought F-8 Crusader

Confusion: Corsair II

Power: 1 × J57 reheated turbojet   *Span:* 10.87m   *Length:* 16.61m

A supersonic carrierborne fighter and reconnaissance aircraft, the Crusader first entered service in 1957 and remains in small numbers with the French Navy. Four 20mm cannon are mounted in the fuselage, and there are four positions for air-to-air missiles, air-to-surface missiles or bombs. Two underwing weapon/fuel tank pylons can also be fitted. The wing has variable incidence for low-speed flight. Versions in use are the F-8H, K, J and L interceptors and the RF-8G reconnaissance aircraft. Performance figures include a maximum level speed of 1,930km/h at 11,000m; service ceiling of 11,700m; and combat radius of 708km. *Country of origin:* USA. *Silhouette and picture:* F-8E(FN).

Power: 1 × TF41 turbofan   *Span.* 11.79m   *Length:* 14.05m

Capable of carrying a massive weapon load, the single-seat subsonic Corsair II is a standard attack aircraft in both the USAF and USN. First flown in September 1965 the A-7 was initially powered by American engines, and was later fitted with the Allison licence-built version of the Rolls-Royce Spey. The USAF version is the A-7D (two seat A7-K), the USN version the A-7E and the two-seat trainer is designated TA-7C. TA-7C modified for fleet operations support group is EA-7L.The type has been exported to Greece and Portugal, the latter being known as A-7P/TA-7P. Up to 6,804kg of bombs, rockets, air-to-surface and air-to-air missiles and gun pods can be carried. A 20mm cannon is mounted in the fuselage. *Country of origin:* USA. *Silhouette:* A-7E. *Picture:* A-7P.

Dassault-Breguet Etendard IVP

Confusion: Hunter, 'Fitter', Super Etendard, G.91

Power: 1 × Atar turbojet   *Span:* 9.6m   *Length:* 14.4m

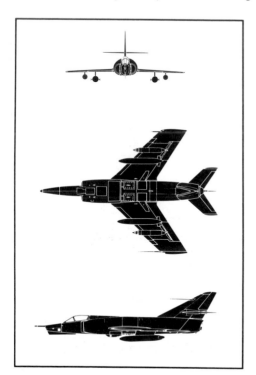

Originally introduced aboard French aircraft carriers in 1962, the Etendard IVM was a single-seat strike fighter. Ninety Etendards were built, of which 21 were designated IVP and equipped with cameras in the nose and gun bay. All the IVPs are fitted with flight-refuelling probes. The IVP remains in squadron service with the Aéronavale, while the IVM has been relegated to training or used in the tanker role as the MP. *Country of origin:* France.

Dassault-Breguet Super Etendard

Power: 1 × Atar turbojet   *Span:* 9.60m   *Length:* 14.31m

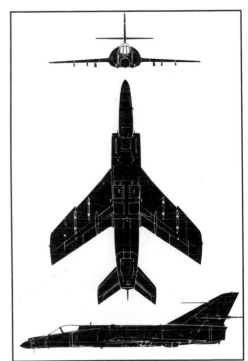

The successor to the Etendard IVM, the Super Etendard is a transonic single-seat strike fighter designed for operation from French aircraft carriers. The engine has increased thrust, a modern navigation/attack system is fitted and an Agave radar is housed in the nose. The wing has also been modified. Production deliveries began in 1978. A total of 71 have been acquired by the French Navy, plus 14 for the Argentinian Navy. As on the IVM, the gun armament consists of twin 30mm cannon. A variety of weapons, including Exocet anti-ship missiles, can be carried on four underwing and one fuselage pylon. A shore-based ground attack version is on offer. *Country of origin:* France.

Tupolev Tu-28P 'Fiddler'

Confusion: 'Fitter'

Power: 2 × Lyulka reheated turbojets   *Span:* 19.8m   *Length:* 27.43m

A long-range interceptor of massive size, the Tu-28P, code-named 'Fiddler', carries two crew, a large radar and up to four air-to-air missiles. As on many Soviet types, the main undercarriage retracts into blisters on the wing trailing edge. Although production is believed to have ceased in the late 1960s, 'Fiddler' remains in Soviet Air Force service in specialist roles. Maximum speed is Mach 1.75 and range 4,989km. *Country of origin:* USSR

Grumman A-6 Intruder/Prowler

Power: 2 × J52 turbojets   *Span:* 16.15m   *Length:* (A-6E): 16.64m, (EA-6B): 18.11m

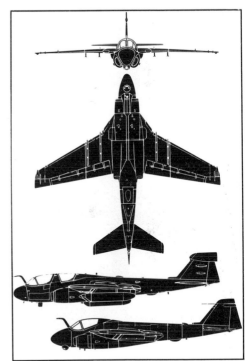

A highly versatile carrierborne aircraft, the Intruder was designed in the late 1950s as a two-seat, low-level, long-range attack bomber for the US Navy, being broadly equivalent to the Royal Navy's contemporary Buccaneer. In addition to its all-weather bombing role (A6A/B/E), the Intruder evolved during the Vietnam War as a tanker (KA-6D) and electronic warfare aircraft, the EA-6A/B Prowler (upper side view). The last-mentioned differs from other models in having a large electronics pod mounted on the fin, while the EA-6B has accommodation for two additional EW operators in an extended nose. An updated version bears the designation A-6F. *Country of origin:* USA. *Main silhouette:* A-6E; *upper side view and picture:* EA-6B.

Power: 1 × JT15D turbofan   *Span:* 8.43m   *Length:* 9.31m

A low-cost lightweight jet basic trainer, the S.211 was developed as a private venture in Italy. The crew are seated in tandem and the four underwing hardpoints can carry bombs or rocket projectiles for weapon training. Maximum speed of the S.211 is 723km/h while maximum cruising range, with external tanks, is 2,693km. The prototype S.211 flew in April 1981 and deliveries to customers, in Singapore and Haiti, began in 1984. *Country of origin:* Italy.

Power: **2 × F3-IHI turbofans**   Span: **9.94m**   Length: **13m**

A tandem two-seat intermediate jet trainer, the Kawasaki T-4 is to replace ageing TI-A/B and T-33 aircraft in the Japanese Air Self Defence Force. First flown in July 1985 as the XT-4, about 200 of the type are due to be built. The first 12 production machines have already been delivered. Various weapons can be carried on four underwing pylons. The T-4 has a maximum speed of Mach 0.9 and a range of 1,668km. *Country of origin:* **Japan.** *Silhouette and picture:* Kawasaki XT-4.

BAe Hawk T1/T-45A Goshawk

Confusion: Hunter, Etendard, Alpha Jet, S.211

Power: 1 × Adour turbofan   *Span:* 9.40m   *Length* (over probe): 11.96m

First flown in 1974, the Hawk T1 was ordered by the RAF as its standard advanced trainer, entering service in 1976. A 30mm Aden gun pack is carried and bombs, guided weapons and rocket projectiles can be carried on the wing pylons. The Hawk 100 systems management trainer has an extended front fuselage. A joint development by McDonnell Douglas and BAe, the T-45A Goshawk is a naval version for use by the US Navy. The undercarriage has been changed, a deck hook fitted and the twin underfuselage strakes replaced by a single unit. Export orders for the basic Hawk total 370 and production of 300 Goshawks is planned. *Country of origin:* UK. *Silhouette:* Hawk T1; *lower side view:* T-45A. *Picture:* T-45A.

Power: 1 × Adour turbofan   *Span:* 9.39m   *Length:* 11.38m

Developed from the Hawk two-seat trainer, the Hawk 200 is a single-seat combat aircraft for use in the air-defence, close air support, photographic reconnaissance and anti-shipping roles. A variety of weapons can be carried, including Sidewinder air-air and Sea Eagle anti-ship missiles. One or two 25mm cannon are mounted in the fuselage. Maximum range, with drop tanks, is 3,610km. Sixty Hawk 200s have been ordered by Saudi Arabia, while the first demonstrator flew in May 1986. *Country of origin:* UK.

SOKO G-4 Super Galeb

Confusion: Hawk, S-211

Power: 1 × Viper turbojet   *Span:* 9.88m   *Length overall:* 11.86m

Designed to replace the Galeb and T-33 as the Yugoslav Air Force basic/advanced trainer, the two-seat Super Galeb (Seagull) first flew in July 1978. Powered by an uprated Viper engine, the Super Galeb can carry a 23mm cannon in a removable ventral pod and has four underwing stores points for rockets, bombs, fuel tanks etc. The aircraft was produced to meet major Yugoslav orders. Maximum speed is 910km/h and combat radius in the light attack role, at low level, is 300km. *Country of origin:* Yugoslavia.

Power: 1 × Orpheus turbojet   *Span:* 6.73m (trainer)   *Length:* 10m (trainer)

Designed originally by the Folland company as an ultra-lightweight interceptor, the Gnat Mk 1 was adopted in that role by India and Finland and as a two-seat trainer by the RAF. Two hundred Gnat Mk 1 fighters were built at Bangalore by Hindustan Aeronautics. Eighty Ajeet ground-attack fighters, with more power and systems improvements, have been built. The Ajeet has also been constructed in two-seat trainer form. Ajeet carries two 30mm cannon plus underwing bombs or rockets. Performance figures for the Gnat/Ajeet include maximum speeds of 1,118km/h at 6,100m and 1,040km/h at 11,000m; ceiling of 15,250m; and combat radius of 805km. RAF Gnats were replaced by the Hawk. *Countries of origin:* India/UK. *Main Silhouette:* Ajeet; *lower side view:* Ajeet trainer. *Picture:* Ajeet.

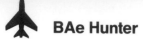

BAe Hunter

Confusion: Ajeet, Hawk, Etendard

Power: 1 × Avon turbojet *Span:* 10.26m *Length:* 13.98m

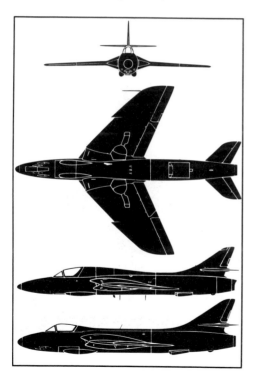

Undoubtedly the most successful and most numerous of post-war British fighter designs, with around 2,000 produced, the Hunter has been largely phased out of service by Western air forces, except in Switzerland and the UK. Many remain in service in the Middle and Far East, Africa, India and South America, however. The RAF still uses Hunters in a variety of roles, including advanced training with the two-seat T7/T8. Hunters carry two internally mounted 30mm guns plus underwing rockets, bombs and fuel tanks. *Country of origin:* UK. *Main silhouette:* FGA9; *upper side view:* T7. *Picture:* T8.

Power: 2 × **Spey turbojets**   *Span:* **13.4m**   *Length:* **19.33m**

Originally built as a Royal Navy low-level, two-seat carrierborne strike/attack aircraft, the Buccaneer was later adopted by the RAF as a land-based type for the same purpose. First production aircraft (S1) had the Gyron Junior engine, but subsequent variants (S2) were powered by the Rolls-Royce Spey. The RAF is now the sole UK Buccaneer operator and the type remains in service, in the maritime role. The South African Air Force has half a dozen Buccaneers, designated S500. Armament is carried internally and on four wing pylons. Bullpup, Sea Eagle or Martel missiles can be carried. Buccaneer performance is characterised by outstanding stability at low level and high speed. *Country of origin:* UK. *Silhouette and picture:* S2.

Saab 105

Confusion: Buccaneer

Power: 2 × Aubisque or 2 × J85 turbojets   *Span:* 9.5m   *Length:* 10.5m

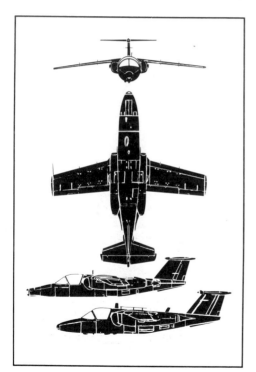

Produced in the 1960s as a standard basic trainer for the Royal Swedish Air Force, the two-seat 105 is unusual in having a shoulder wing and T-tail. Four versions have been built: the Sk60A for training, Sk60B for weapons training/light attack, Sk60C for reconnaissance (all with Turboméca Aubisque engines), and the 105XT exported to Austria with General Electric J85 engines. The Sk60C (upper side view) has a longer nose than other variants. Some one hundred and sixty 105s are in service. *Country of origin:* Sweden. *Main silhouette and picture:* Sk60A.

Power: 2 × F100 reheated turbojets   *Span:* 13.04m   *Length:* 19.44m

A sophisticated single-seat, all-weather air-superiority fighter, the F-15 Eagle has a performance in excess of Mach 2 at altitude. In addition to a rotary cannon in the fuselage, the F-15 can carry four Sparrow and four Sidewinder air-to-air missiles. Operating in the USA and Europe with the USAF, the F-15 has also been sold to Israel, Japan and Saudi Arabia. First production aircraft flew in 1974. Current production single-seaters are F-15C and F-15D. A dual role two-seater (air intercept and ground attack) is the F-15E of which 392 are scheduled to be built. The tandem two-seat version is designated F-15B. *Country of origin:* USA. *Silhouette:* F-15. *Picture:* F-15E.

Mikoyan MiG-25 'Foxbat'

Confusion: Tomcat, Eagle, 'Foxhound'

Power: 2 × Tumansky reheated turbojets   *Span:* 12.5m   *Length:* 21.33m

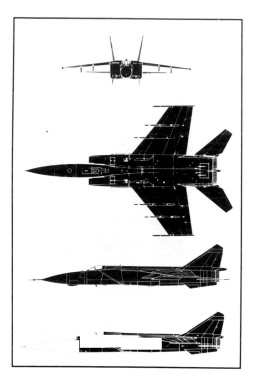

A high-performance, high-altitude interceptor, the MiG-25 'Foxbat' is capable of speeds of between Mach 2.5 and 3.0. Because of its very high speed and 24,400m operational ceiling 'Foxbat' presents a problem to any air-defence system. 'Foxbat A' is an interceptor with a large nose radar and four underwing air-to-air missiles. 'Foxbat B' and 'D' are reconnaissance variants. Foxbat C (MiG-25U) is a two-seat trainer and 'Foxbat E' is a development of the 'A' with new radar and uprated engines. 'Foxbat F' is believed to be an electronic warfare variant. *Country of origin:* USSR. *Main silhouette:* 'Foxbat A'; *lower side view:* 'Foxbat C'. *Picture:* 'Foxbat E'.

Mikoyan MiG-31 'Foxhound'

Power: 2 × Tumansky reheated turbojets   *Span:* 14m   *Length:* 21.5m

A two-seat development of the 'Foxbat', the MiG-31 'Foxhound' Mach 2.4 interceptor has a new radar, modified fuselage and revised wing shape. Having a full look down/shoot down capability, the type is believed to have entered service in 1983 and is now operational in strength. Four AA-9 Amos radar homing long range air-air missiles can be carried under the fuselage. A further four AA-9s or smaller weapons can be mounted on wing pylons. Soviet sources claim that 'Foxhound' is capable of intercepting cruise missiles at low altitude. *Country of origin:* USSR.

Mikoyan MiG-29 'Fulcrum'

Confusion: F-15, 'Flanker', Hornet, 'Foxbat'

Power: 2 × Tumansky reheated turbofans   *Span:* 11.5m   *Length (overall):* 17.2m

Similar in size and layout to the F-18 Hornet, the MiG-29, codenamed 'Fulcrum', was first referred to in 1979. It is an all-weather single seat interceptor first deployed in the Soviet Union early in 1985. With two Tumansky engines, the 'Fulcrum' is capable of speeds in excess of Mach 2 and has a combat radius of 1,150km. It carries medium and short range air-air missles, rockets or bombs on six underwing pylons. A 30mm cannon is mounted in the port wingroot. 'Fulcrums' have been supplied to India, Syria and Iraq. *Country of origin:* USSR.

Power: 2 × Lyulka reheated turbofans   *Span:* 14m   *Length.* 21.9m

Approximating to the American F-15, the 'Flanker' is a supersonic all-weather counterair fighter of the same new generation as the MiG-29 'Fulcrum'. Like the 'Fulcrum', the 'Flanker' has twin fins and it can carry a variety of air-air missiles on underfuselage, underwing pylons and wingtips. One 30mm cannon is mounted in the starboard wingroot extension. Maximum speed is estimated at 2,120km/h and combat radius 1,500km. 'Flanker' is mainly intended to combat low flying aircraft and cruise missiles. The 'Flanker B' is in service in growing numbers. The two-seat version is known as 29UB. *Country of origin:* USSR.

Jet, swept wing, flank/root engine(s)

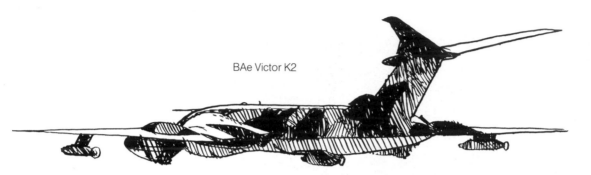

BAe Victor K2

Tupolev Tu 16 'Badger'

Power: 4 × Spey turbofans   *Span:* 35m   *Length:* 38.63m

Developed from the basic Comet airframe, the Nimrod MR2 is the RAF's standard long-range maritime reconnaissance aircraft. Capable of up to 12h endurance and a maximum speed of 926km/h, the Nimrod is one of the most advanced anti-submarine aircraft in the West. Three Nimrod R1s are used for electronic surveillance. The R1 has a cut-off fuselage tail extension and small inset fins on the tailplane. Armament of the maritime reconnaissance variants includes bombs, mines, depth charges, Stingray torpedoes and Sidewinder and Harpoon missiles. Following a recent systems update, the MR2s have wingtip ESM pods, finlets on the tailplane and a small rear underkeel. With an inflight refuelling probe, they are known as MR2P. *Country of origin:* UK. *Silhouette:* MR2. *Picture:* MR2P.

BAe Victor   *Confusion:* 'Badger', 'Bison'

Power: 4 × Conway turbofans   *Span:* 36.58m   *Length:* 35m

One of the three V-bombers which formed the British nuclear deterrent force of the 1950s and 1960s, the Victor was unusual in having a crescent-shaped wing leading edge and high-set tailplane. The B1 had Sapphire engines while the B2 has Rolls-Royce Conways. The Victors in RAF service are all K2 tankers. Three aircraft can be refuelled simultaneously through hoses trailed from the belly and pods under the wings. The Victor is scheduled, later, to be replaced by VC10 and TriStar tankers. *Country of origin:* UK. *Silhouette and picture:* K2.

Power: 2 × Mikulin turbojets   *Span:* 32.93m   *Length:* 34.8m

First put into service in the mid-1950s as a bomber, the 960km/h Tu-16 'Badger' remains in use with Soviet maritime squadrons. Used for overwater surveillance and electronic intelligence-gathering, the 'Badger D' has a long nose radome and three underfuselage blisters. 'Badger E' and 'F' have glazed noses, the 'E' mounting only two blisters and having a belly camera, and 'F' having an electronic equipment pod under each wing. 'Badgers C' and 'D' carry 'Kingfisher' air-to-surface missiles under the wings or a 'Kipper' missile under the fuselage. 'Badgers H' and 'J' are used for ECM work. 'Badger K' is an electronic reconnaissance variant. 'Badgers' use air-to-air refuelling and some act as tankers. There are twin tail guns. Maximum range with the maximum bomb load of 9,000kg is 4,800km, increasing to 6,400km with 3,000kg at 770km/h. *Country of origin:* USSR. *Silhouette:* 'Badger C'. *Picture:* 'Badger D'.

Myasishchev M-4 'Bison'   *Confusion:* 'Badger'

Power: 4 × Soloviev turbofans   *Span:* 51.82m   *Length:* 49.38m

Originally designed as the rival to the Boeing B-52 strategic bomber, the M-4, code-named 'Bison', first flew in 1953 and entered service with the Soviet Air Force in 1955. Initial versions were powered by four Mikulin turbojets but later machines were fitted with Soloviev turbofans. Later the 'Bison' was converted to the 'B' and 'C' standards for the long-range maritime reconnaissance role. There are three positions for twin 23mm cannon. Some 'Bisons' are used as airborne refuelling tankers. A number of radar and electronic warfare sets are carried, and a special feature is the bicycle undercarriage with two sets of mainwheels under the fuselage and outrigger wheels at the wingtips. Maximum speed is 1,060km/h and endurance is about 15h. *Country of origin:* USSR. *Silhouette:* 'Bison A'. *Picture:* 'Bison B'.

An-225

Il-96-300

Lockheed Tristar

Lockheed S-3A/3B 'Viking'

Confusion: Skywarrior, C-1A

Power: 2 × TF34 turbofans   *Span:* 20.93m   *Length:* 16.26m

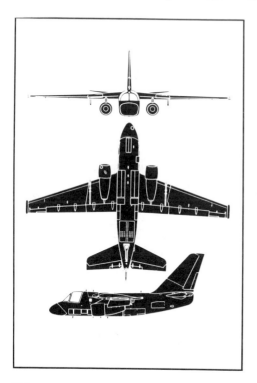

A highly sophisticated carrierborne anti-submarine aircraft, the S-3A Viking has podded turbofans and top speed of 816km/h. The fuselage weapons bay carries torpedoes, depth charges or mines, while weapons or fuel tanks can be carried on wing pylons. Four crew members are carried. A carrier on-board delivery (COD) version is designated US-3A and an inflight refuelling type, the KS-3A. The prototype of the US-3A, produced by modifying a development S-3A, first flew in 1976 and underwent sea trials aboard the USS *Kitty Hawk* in 1978. S-3As are being modified as S-3Bs with new radar, improved electronic systems and Harpoon anti-ship missiles. *Country of origin:* USA. *Silhouette:* S-3A. *Picture:* S-3B.

Power: 2 × JT8D turbofans   *Span:* 30.6m   *Length:* 29m

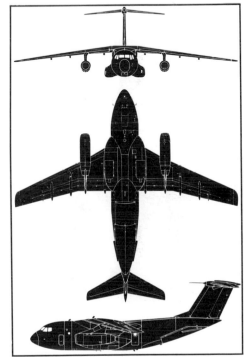

A medium-range tactical transport, the C-1A equips the Japan Air Self-Defence Force. Up to 60 troops or vehicles/freight, loaded via the rear doors and ramp, can be carried. Twenty-four C-1As were delivered, in addition to two prototypes and two pre-production models. Maximum speed is 787km/h, and range with full payload 1,295km. At reduced payload 2,200kg and with maximum fuel, range increases to 3,353km. One C-1, designated C-1 Kai, has been fitted with larger flat radomes in nose and tail for ECM purposes. *Country of origin:* Japan.

115

Yakovlev Yak-28P 'Firebar'

Confusion: 'Flashlight', 'Brewer'

Power: 2 × Tumansky reheated turbojets   *Span:* 12.5m   *Length:* 21.95m

Developed from the 'Flashlight/Mangrove' series, the Yak-28P 'Firebar' is an all-weather interceptor with low-supersonic performance. 'Firebar' has a long, pointed nose radome, two seats in tandem and two or four underwing air-to-air missiles. The 'Firebar' first flew in 1960, was built in large numbers and remains in limited service with the Eastern Bloc. Like the rest of the family, the 'Firebar' has a bicycle undercarriage with outrigger wheels near the wingtips. At high altitude the 'Firebar' has a radius of action of 885km. *Country of origin:* USSR.

Power: 2 × Tumansky reheated turbojets　　*Span:* 12.05m　　*Length:* 21.34m

A redesigned and higher-powered successor to the 'Flashlight', the 'Brewer' has a glazed nose similar to that of an interim reconnaissance variant, the Yak-27R 'Mangrove', which differs little externally. 'Brewer' is a reconnaissance 'D' and electronic countermeasures 'E' aircraft. A key task of 'Brewers' now in Soviet service is electronic warfare, for which a variety of sensors are housed in fuselage blisters. The trainer version is the Yak-28U, code-named 'Maestro', which has an additional cockpit forward of the main position and no glazing on the nose. *Country of origin:* USSR. *Silhouette and picture:* 'Brewer D'.

Douglas A-3 Skywarrior

Confusion: Viking, C-1A

Power: 2 × J57 turbojets   *Span:* 22.10m   *Length:* 23.27m

Introduced in the 1950s and originally used as a standard US Navy bomber, the Skywarrior now performs the electronic countermeasures and tanker roles. The ECM version, with a variety of fuselage bulges housing special equipment, is known as the EKA-3B. The airborne tanker is designated KA-3B. The original armament of two 20mm cannon in the tail has been deleted from the present versions, which are conversions of earlier bomber/reconnaissance aircraft. Some 50 Skywarriors are still in service. *Country of origin:* USA. *Silhouette and picture:* EKA-3B.

Power: 2 × JT8D turbofans   *Span:* 28.35m   *Length:* 30.48m

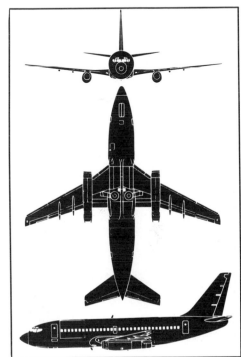

A very successful short/medium-haul transport, this type first entered service with Lufthansa in 1967 as the 115-seat 737-100. Seating was subsequently increased to 130 in the 737-200, a variant which has sold in large numbers round the world. Nineteen 737s designated T-43A were delivered to the USAF as navigation trainers. The 737-200 can fly 3,815km with maximum payload. Maximum cruising speed is 927km/h. *Country of origin:* USA. *Silhouette and picture:* 737-200.

Boeing 737-300/400

Confusion: 737-200, A310, 767

Power: 2 × CFM56 turbofans   *Span:* 28.88m   *Length:* (-300): 33.4m, (-400): 36.45m

The 737-300 and -400 differ from the -200 in wing and tailplane span, fuselage length, fin outline and engine nacelle shape. The -300 seats up to 149 passengers and the -400, up to 170. Fuselage length is increased by 3.05m on the -400. The -300 first flew in February 1984 and the -400 in February 1988. By May 1989, 791 firm orders had been announced for the 737-300 and 156 for the -400. With 141 passengers, at normal weight, the -400 has a range of 2,993km. *Country of origin:* USA. *Silhouette:* -300. *Picture:* -400.

Boeing 737-500

Power: 2 × CFM56 turbofans   *Span:* 28.88m   *Length:* 36.45m

A combination of 737-300 and -400 technology and a shortened fuselage, the -500 accommodates 108 passengers and has, with auxiliary fuel tanks, a maximum range of 5,552km. The -500, in effect, replaces the earlier -200. Firm, announced, orders by May 1989, totalled 132 and first deliveries were due in March 1990. *Country of origin:* USA.

Airbus A320

Power: 2 × CFM56 or 2 × V2500 turbofans   *Span:* 33.91m   *Length:* 35.75m

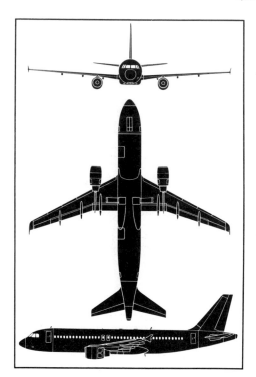

Using advanced design techniques, materials and avionics, the Airbus A320 represents a new class of short/medium range twin jet airliners. First flown in February 1987, the 150-179 seat A320 is built jointly by Aérospatiale in France, MBB (Deutsche Airbus) in Germany, British Aerospace PLC in the United Kingdom, Belairbus in Belgium and CASA in Spain. At the time of writing the order book stood at 451 aircraft for 24 customers. The A320 has a fly-by-wire control system and side stick controllers in place of the usual control columns in the cockpit. Range with CFM56 engines and 150 passengers is 5,318km. *Countries of origin:* France/Germany/Belgium/Spain/UK plus other associated countries.

Power: 2 × CF6 or 2 × JT9D turbofans   *Span:* 44.84m   *Length:* 53.62m

First flown in 1972, the 330-seat wide-body A300 is built by a European consortium of companies: Aérospatiale, MBB (Deutsche Airbus), British Aerospace PLC, CASA and Fokker. The B2 and B4 are basic versions, the latter having more fuel and leading-edge flaps. The A300 is intended for medium-range routes and can fly 4,074km with full payload and reserves. Maximum cruising speed is 930km/h. Two further versions are the A300-600 and A300-600R with increased passenger and freight capacity. The -600R has extended range while the -600 convertible has mixed passenger/freight capacity. By March 1989, the order book for A300s stood at 336. *Countries of origin:* France/Germany/Netherlands/Spain/UK, plus other associated countries. *Silhouette:* A300B2. *Picture:* A300-600.

Airbus A310

Confusion: A300, 757, 767, TriStar, DC-10

Power: 2 × CF6 or JT9D turbofans   *Span:* 43.9m   *Length:* 46.6m

A combination of a shortened-fuselage A300 and a new advanced-technology wing produced the Airbus A310, a direct rival to the Boeing 767. Like the 767, the A310 is offered with two different powerplants. By March 1989 180 A310s had been ordered. First flown in April 1982, the type went into service in April 1983, with Lufthansa and Swissair the first two users. With maximum payload the A310 can fly 4,090km at a speed of 828km/h. A310-200 is the basic version, the -200C is a convertible version, and -200F a freighter. A first delivery of the A310-300, with longer range, took place in April 1986. Wingtip fences were introduced in 1986. *Countries of origin:* France/Germany/Netherlands/Spain/UK, plus other associated countries. *Silhouette:* A310-200. *Picture:* A310-300.

Power: 3 × RB 211 turbofans   *Span:* 47.35m   *Length.* (except -500): 54.17m

First flown in September 1981, the Boeing 767 is a medium-haul wide-body airliner seating a maximum of 255 passengers. The 767 and the Airbus A310 are direct competitors and they closely resemble each other in external shape. First delivery of the 767 was made in August 1982. By March 1989 firm orders totalled 358. Like its sister aircraft the 757, the 767 is offered with a choice of powerplants. The 767-200ER has increased range, the 767-300 has a fuselage extended by 6.42m, while the 767-300ER has range increased to over 10,600km. *Country of origin:* USA. *Silhouette:* 767-200. *Picture:* 767-300.

125

Boeing 757   *Confusion:* A300, A310, 767

Power: 2 × RB 211 or 2 × PW2037 turbofans   *Span:* 37.95m   *Length:* 47.32m

Reversing the trend towards wide-body twin-aisle transports, the slim-body twin-fan single-aisle Boeing 757 airliner first flew in February 1982, with deliveries to airlines beginning ten months later. Firm orders for the 757 totalled 564 by June 1989, with some airlines selecting Rolls-Royce and others Pratt & Whitney powerplants. In mixed-class international layout the 757 seats 204 passengers, cruises at 850km/h and has a maximum range of over 3,984km. The 757-200PF is a freighter with a large side cargo door, while the 757-200M is a mixed passenger/cargo version. *Country of origin:* USA. *Picture:* 757-200.

Power: 3 × RB211 turbofans   *Span:* 47.35m   *Length:* (except -500): 54.17m

Produced to meet the same wide-body requirement as the McDonnell Douglas DC-10, the Lockheed TriStar was the first aircraft to be equipped with the Rolls-Royce RB211 big-fan engine. The various versions are the 1011-1, 1011-100, 1011-200, 1011-250 and the 1011-500; the last-named has a 4.11m reduction in fuselage length. Capacity varies, with up to 400 passengers in all-economy. Lockheed closed the TriStar line in August 1983 with the 250th production aircraft. Nine ex-airline 1011-500s are being converted for the RAF by Marshall of Cambridge, seven as tanker/passenger aircraft (TriStar K1 and K2) and two as tanker/freighters (KC1). *Country of origin:* USA. *Silhouette:* L-1011-500. *Picture:* TriStar K1.

McDonnell Douglas DC-10   *Confusion:* TriStar, A300, A310, 767

Power: 3 × CF6 or 3 × JT9D turbofans   *Span:* (30/40): 50.41m   *Length:* (30/40): 55.5m

Over 386 DC-10 wide-body airliners have been sold to 53 operators. The Series 10 was the initial version, followed by the Series 30 and 40, both with 3.05m more wingspan. The Series 10 and 30 have CF6 engines and the Series 40 the JT9D. Convertible cargo versions are in service and the USAF has ordered the type for use as an airborne tanker/cargo aircraft designated KC-10. Like the TriStar, which it resembles, the DC-10 first flew in 1970. Maximum seating is for 380 and maximum range is 11,580km. *Country of origin:* USA. *Silhouette:* DC-10-30. *Picture:* KC-10A.

Power: 3 × PW 4360 or RB211 or CF6 engines   *Span:* 51.66m   *Length:* 61.21m

Developed from the DC-10, the MD-11 has a longer fuselage, re-shaped tailplane and winglets above and below the wingtips. The standard MD-11 will seat from 276 to 405 passengers and maximum range is 12,985km. The MD-11ER (extended range) has the same length fuselage as the DC-10 and has a maximum range of 14,594km. The MD-11F is all-freight. Economical cruising speed is 876km/h. The company is also planning a 'Stretch' version with 90 additional passengers in a fuselage plug and 'Superstretch' with a larger wing and seating for up to 612 passengers. Scheduled to enter service in 1990, the MD-11 had, by March 1989, amassed a total of 253 orders and commitments. *Country of origin:* USA. *Picture:* artists impression.

McDonnell Douglas DC-8

Confusion: 707, Convair 880/990

Power: 4 × JT3D turbofans   *Span:* 43.41m   *Length:* 45.87m

Along with the Boeing 707 and the Comet, the DC-8 pioneered the airline jet age, and 294 of the Series 10-50 were built. First flown in 1958, the DC-8 went into service in 1959, some fitted with Rolls-Royce Conway engines. The Series 50 can carry up to 173 passengers for 9,205km. Known as the Jet Trader, a freight and freight/passenger version was introduced into airline service in 1962. *Country of origin:* USA. *Silhouette:* DC-8-50. *Picture:* DC-8-53.

McDonnell Douglas DC-8 Super 60/Super 70

Power: 4 × JT3D turbofans   *Span:* (Super 63): 45.23m   *Length* (Super 63): 57.12m

The process of 'stretching' jet transports is nowhere more evident than in the DC-8, the ultimate variant of which, the Series 63, has grown by more than 9.14m compared with earlier versions. The Super 62, with more wingspan than the Series 50, is the shortest version of this range. The Super 61 has the Series 50 wing and a 6.08m fuselage extension, while the Super 63 has the longest fuselage and the bigger wing. The Super 63 can carry up to 259 passengers. In all, 262 Super Sixties were built. Re-engined with CFM56 advanced turbofans, the 61, 62 and 63 became the Super 71, 72 and 73 respectively. Some seventy of the Super 70 series are now in service. *Country of origin:* USA. *Silhouette:* Super 63; *lower side view:* Super 73. *Picture:* Super 70.

131

Boeing 707

Confusion: DC-8, KC-135, A340

Power: 4 × JT3D turbofans   *Span:* 44.2m   *Length:* 46.61m

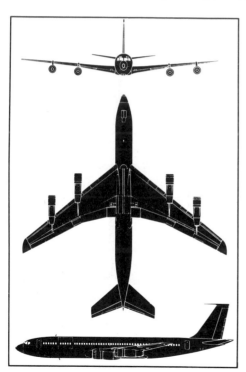

The most successful of the first generation of medium/long-range jet airliners, the 707 first flew in prototype form in 1954. Initial versions seated up to 181 passengers , while the smaller 720 seated 153. Later versions such as the -320 and -420 carry up to 195 passengers and there are variants for cargo and mixed passenger/cargo. The 707 has a maximum cruising speed of 973km/h and can fly 10,040km with maximum fuel. Boeing offers conversions of the 707 as a mixed tanker/transport with hose-drogue or boom refuelling equipment. A total of 982 707/720s have been produced. *Country of origin:* USA. *Silhouette:* 707-320. *Picture:* 707-320C.

Confusion: 707, DC-8 and A340   **Boeing KC-135**

Power: 4 × J-57 turbojets, or JT3D or CFM56 turbofans   *Span:* 39.88m   *Length:* 41.53m

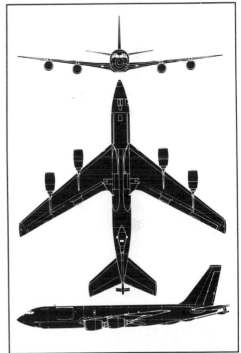

First flown in 1956, the KC-135A is an air-air refuelling tanker version of the 707 airliner. Altogether 732 KC-135As with J57 engines were built. To extend the life of the type, JT3D turbofans are being fitted to some aircraft, while 300 others, plus 11 French Air Force machines, are re-equipped with General Electric/SNECMA CFM56 turbofans with much more bulbous nacelles. This version is known as the KC-135R. *Country of origin:* USA. *Silhouette:* KC-135R. *Picture:* KC-135 with JT-3D.

Boeing E-3A Sentry

Confusion: DC-8, 707, 'Moss'

Power: 4 × TF33 or CFM56 turbofans   *Span:* 44.42m   *Length:* 46.61m

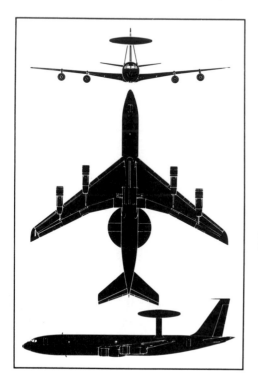

A radar station in the sky, the E-3A Sentry is also known as AWACS (Airborne Warning and Control System). The aircraft is basically a 707 transport fitted with a very large circular radome above the fuselage and a complete air-defence operations centre in the cabin. The 17-man crew track numerous hostile targets and direct fighters onto them. The Sentry has been produced for the US Air Force, NATO, the RAF and Saudi Arabia. The first development aircraft, flown in 1972, was known as the EC-137. The type is fully operational in the USA and Europe and sub-variants are E-3B and E-3C. *Country of origin:* USA. *Silhouette and picture:* E-3A.

Power: 4 × Kuznetsov turbofans   *Span:* 48.06m   *Length:* 59.54m

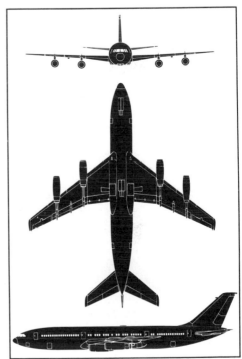

The Soviet Union's first wide-bodied jet, the Il-86, Nato code-named 'Camber', first reported in 1972, subsequently underwent extensive redesign, and finally flew in December 1976. Seating up to 350 passengers, the Il-86 is intended for use over sectors of 2,350km at a speed of around 900–950km/h. The Il-86 entered service with Aeroflot in 1980. To improve performance, the Il-86 may be re-engined with Soloviev turbofans. *Country of origin:* USSR.

Ilyushin Il-96-300

Confusion: Il-86, 747-400

Power: 4 × PS-90A turbofans   *Span:* 57.66m   *Length:* 55.35m

Resembling the Il-86, the Il-96-300 is, in fact, a new design with an increased span wing of less sweep back and fitted with tip winglets. The 96-300 has Soloviev instead of Kuznetsov turbofans and passenger loading will be through the upper and not lower deck as on the Il-86. Accommodation is for 250–300 passengers, maximum range is 11,000km and cruising speed 850-900km/h. The first prototype has flown and production has begun. *Country of origin:* USSR.

Power: 4 × JT9D, CF6 or RB 211 turbofans   *Span:* 59.64m   *Length:* 70.51m

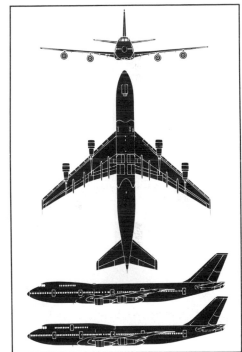

The Boeing 747 was the first of the wide-bodied 'jumbo jets'. It can seat up to 490 passengers in economy class and cruise at 948km/h over sectors of up to 11,395km. Initial version was the -100, followed by the -200 with more fuel. Passenger/cargo and all-cargo 747s are known as the -200C and -200F respectively. A short-range variant is designated 747SR. The short-fuselage 747SP (Special Performance) carries 360 passengers for up to 15,400km. Latest version is the 747-300 (formerly 747EUD), with the upper front fuselage extended aft to accommodate 37 more passengers. Nineteen PanAmerican 747s are being modified to passenger/convertible standard as C-19As as part of the Civil Reserve Air Fleet programme of the USAF. *Country of origin:* USA. *Main silhouette:* 747-200B; *lower side view:* 747-300. *Picture:* 747-300.

Boeing 747-400   *Confusion:* A340, MD-11

Power: 4 × CF6, PW4256 or RB 211 engines   *Span:* 64.92m   *Length:* 70.66m

A very long range advanced development of the 747-300, the 747-400 has wingspan extended by 3.66m and sweptback winglets fitted at the tips. The -300 fuselage extended upper deck is retained. Engine nacelles and pylons have been standardised with those of the 767. Depending on layout, the 747-400 will carry between 450 and 630 passengers. With 412 passengers the -400 has a range of 13,528km. By February 1989 the -400 order book totalled 192. A combined passenger/freight version is designated 747-400M. *Country of origin:* USA.

Power: 8 × J57 turbojets or 8 × TF33 turbofans   *Span:* 56.42m   *Length:* 48.03m

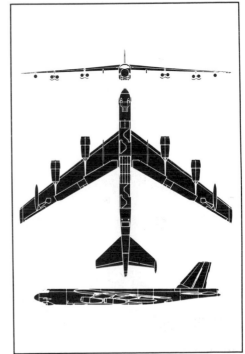

In service since 1957, the B-52 still forms the backbone of the USAF's Strategic Air Command with nearly 300 available. The force is being progressively modified to carry the Boeing AGM-86B air-launched cruise missiles. Main versions in service are the B-52G and B-52H, the former with J57 and the latter with TF33 engines. With a crew of six and full weapon load, the B-52 can carry 27,215kg of internal weapons, plus missiles or other equipment on wing pylons. Outboard external fuel tanks can be fitted. The aircraft carries a wide variety of ECM systems. A multi-barrel radar-controlled 20mm Gatling gun is carried in the tail. *Country of origin:* USA. *Silhouette and picture:* B-52H.

Lockheed C-141B Starlifter

Confusion: Galaxy, 'Candid', BAe 146

Power: 4 × TF33 turbofans   *Span:* 48.74m   *Length:* 51.29m

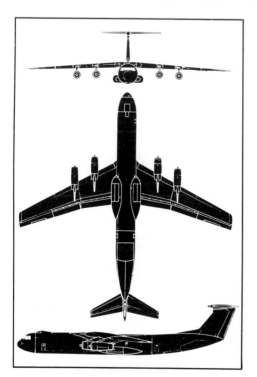

The USAF's first pure-jet strategic freighter, the Starlifter entered service in 1964. Altogether 285 Starlifters were built and they are in use in many parts of the world with Military Airlift Command. Cargo, vehicles, equipment, guns and missiles can be carried. Range with maximum payload of 44,450kg is 4,264km. The Starlifter is loaded through rear under-fuselage doors. All Starlifters in the USAF have now been rebuilt to C-141B standard, with longer fuselage. *Country of origin:* USA. *Silhouette and picture:* C-141B.

Power: 4 × D-30 turbofans   *Span:* 50.5m   *Length:* 46.59m

The Soviet answer to the Lockheed Starlifter, the Il-76 first flew in 1971. A massive strategic transport, the 'Candid' is designed to operate from short, rough airfields. Maximum payload is 40,000kg and cruising speed 850km/h. Tanks, guns and a variety of other vehicles and equipment can be loaded through rear doors. There are two versions, 'Candid A' and 'B', the 'B' being a military version with a tail turret. Several hundred Il-76s are in Soviet Service and the type has been exported to Czechoslovakia, Poland, Iraq, India, Syria, Libya and Cuba. An airborne tanker version has the NATO code-name 'Midas'. *Country of origin:* USSR.

IL-76 'Mainstay'   *Confusion:* E-3A, 'Candid'

Power: 4 × Soloviev turbofans   *Span:* 50.5m   *Length:* 46.59m

A direct development of the IL-76 'Candid' transport, the 'Mainstay' is the Soviet opposite number to the Boeing E-3 Sentry as an airborne early warning and control aircraft. Like the E-3 it has a large rotodome mounted on struts on the upper rear fuselage. The dorsal fin has a prominent air intake at the forward end. Some 15 'Mainstays' are believed to be in service. *Country of origin:* USSR.

Power: 4 × **TF39 turbofans**   *Span:* 67.88m   *Length:* 75.54m

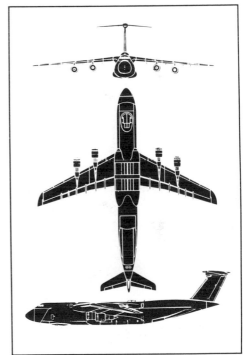

The world's third largest transport, the Galaxy can carry up to 345 troops on two decks, or 120,200kg of freight for 4,745km. First flown in 1968, the Galaxy went into service as the C-5A in 1970. The 76 aircraft in service form four squadrons in Military Airlift Command. Cruising at 864km/h, the Galaxy has a maximum-fuel range of 10,460km. All Galaxies have been retrofitted with a new, stronger wing. The USAF has been receiving an additional 50 aircraft, designated C-5B, an up-dated version which is externally similar. *Country of origin:* USA.

Antonov An-124 'Condor'

Confusion: An-225, Galaxy

Power: 4 × D-18T turbofans   *Span:* 73.30m   *Length:* 69.10m

Until the advent of the An-225, the An-124, NATO code-name 'Condor', was the world's largest aircraft. The prototype 'Condor' transport first flew on 26 December 1982. With an upward opening hinged nose and rear ramp/door, the 'Condor' can be loaded from both ends. The aircraft has a maximum payload of 150,000kg, maximum cruising speed of 865km/h and range with maximum fuel of 16,500km. The 'Condor' carries a flight crew of six, a relief crew of six and can seat 88 passengers in a cabin above the flight deck. It is reported that up to 10 'Condors' per year are to be produced. *Country of origin:* USSR.

Power: 6 × D-18T turbofans   *Span:* 87m   *Length:* 84m

The world's largest aircraft, the six-engined An-225 Mriya (Dream) made its maiden flight in December 1988 from Kiev. Developed from the An-124 'Condor', the An-225 is designed to carry big piggyback loads such as the Soviet space shuttle. Large loads can also be carried in the 43 metre long cargo hold. Maximum take-off weight is 600,000kg and payload 200,000kg. The aircraft has four nosewheels and 14 mainwheels on each side of the fuselage. Apart from sheer size and number of engines, the An-225 differs from the An-124 in having twin fins to facilitate loading on top of the fuselage. *Country of origin:* USSR.

BAe 146 *Confusion:* Starlifter, 'Candid'

Power: 4 × ALF502 turbofans   *Span:* 26.34m   *Length:* (Series 100): 26.16m

Unusual as a civil transport, combining four jets, high wing and T-tail, the BAe 146 is a short-field, short-haul feederliner first flown in September 1981. Three basic versions are produced: the Series 100 with 71/93 seats, and the Series 200 with length increased to 28.55m and seating 82/109 and the series 300 with length extended by 4.8m and seating 120. Mixed passenger/freight versions are on offer and the 146 STA (Small Tactical Airlifter) for military use. The Series 200 cruises at 702km/h over ranges of up to 2,743km. The BAe 146 order book so far covers 153 firm orders. Two Series 100s were ordered for the Queen's Flight. *Country of origin:* UK. *Main silhouette:* 146 Series 200; *lower side view:* 146 Series 100. *Picture:* Series 300.

Tupolev Tu-95/Tu-142 'Bear'

Power: 4 × Kuznetsov turboprops   *Span:* 48.5m   *Length:* 47.5m

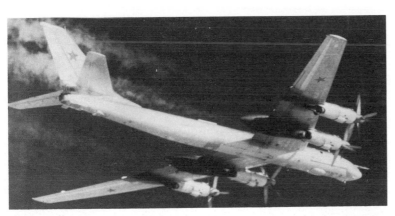

Originally employed by the Soviet Air Force as a long-range strategic bomber 'Bear A' and 'B', the 'Bear' is now used for maritime and electronic warfare work by the Soviet Navy. The 805km/h 'Bear' is very unusual in having swept wings combined with turboprop engines. 'Bear C' has a nose radome and rear-fuselage radomes, 'Bear E' has a smaller nose radome and under-fuselage cameras, and 'Bear F' has larger wing trailing-edge fairings. Standard armament is two 23mm cannon in the tail. Performance figures for 'Bear A' include a maximum level speed at 12,500m of about 805km/h and maximum range of 12,555km. A new version of the aircraft, 'Bear H', is now in production and carries AS-15 cruise missiles. 'Bear J' is a communications relay aircraft. *Country of origin:* USSR. *Main silhouette:* 'Bear B'; *middle side view:* 'Bear C'; *lower side view:* 'Bear D'. *Picture:* 'Bear H'.

Tupolev Tu-126 'Moss'

Confusion: 'Bear', 'Sentry'

Power: 4 × Kuznetsov turboprops   *Span:* 51.2m   *Length:* 57.3m

Adapted from the Tu-114 'Cleat' airliner (which has the wings and tail of 'Bear'), the 'Moss' is an airborne early warning and control aircraft in the same operational class as the Sentry. The 'Moss' carries a massive radar scanner of 11.4m diameter above the rear fuselage. The cabin operations room houses radar operators and control staff. The aircraft is equipped for in-flight refuelling. 'Moss' is designed to work with advanced interceptors. Having located incoming low-level strike aircraft, 'Moss' would pass their height and speed to fighters armed with 'snap-down' missiles capable of being fired from a height of 6,100m or more. It could also warn Soviet strike aircraft if they were about to come under attack. About 7 Tu-126s are in service with the Soviet Air Force. *Country of origin:* USSR.

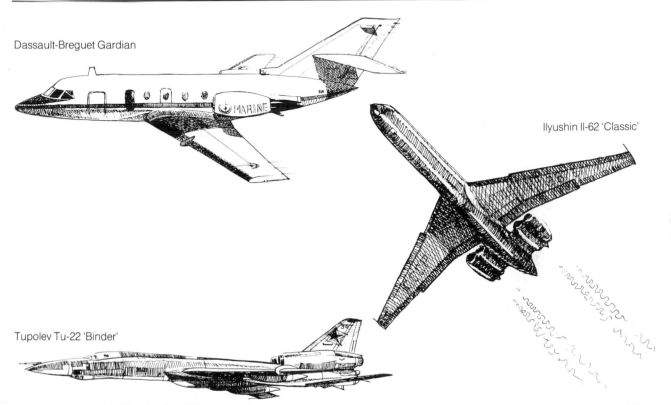

Dassault-Breguet Gardian

Ilyushin Il-62 'Classic'

Tupolev Tu-22 'Binder'

Gulfstream Aerospace Gulfstream II/III/IV

Confusion: One-Eleven, Fellowship, Falcon

Power: 2 × Spey or (Gulfstream IV) Tay turbofans
Length: (II and III): 24.36m; (IV): 26.92m

Span: (II and III): 20.98m; (IV): 23.72m;

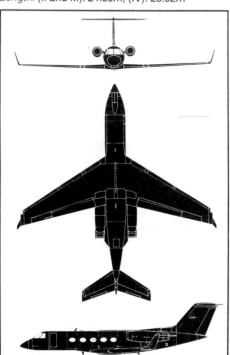

The Gulfstream II is one of the largest purpose-built executive transports, offering luxury accommodation for 19 passengers. The aircraft cruises at 976km/h and range with maximum fuel is an impressive 6,880km. The Gulfstream II first flew in 1966. Wingtip tanks can be fitted. More than 40 Gulfstream IIs were converted to II-Bs, incorporating Gulfstream III wings. The first Gulfstream III with lengthened fuselage and Whitcomb winglets, first flew on 2 December, 1979. The re-engineered, advanced Gulfstream IV, with Rolls-Royce Tay turbofans, first flew in September 1985 and over 100 have been ordered. US military designation for the Gulfstream III is C-20. *Country of origin:* USA. *Silhouette:* Gulfstream III. *Picture:* Gulfstream IV.

Power: 2 × TFE731 turbofans   *Span:* 16.3m   *Length:* 16.9m

The Citation III bears little resemblance to its straight-wing predecessors, described in the next section. With a high-aspect-ratio swept wing and T-tail, the Citation III is an executive transport seating six to thirteen passengers. Cruising speed is 746km/h and range with six passengers is 4,600km. The prototype Citation III first flew in May 1979 and production deliveries began in spring 1983. By early 1988 over 140 Citations were in service. *Country of origin:* USA.

Aérospatiale Caravelle

Confusion: One-Eleven, DC-9

Power: 2 × Avon or 2 × JT8D turbojets   *Span:* (Caravelle 12): 34.30m   *Length:* 36.24m

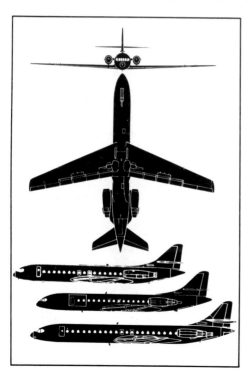

Designed originally by Sud-Est, the Caravelle was the first pure-jet short-haul airliner to go into service. Caravelles I, IA, III, VI-N and VI-R had Rolls-Royce Avon engines while the Series 10R, 11R, Super B and 12 have the Pratt & Whitney JT8D. Early versions accommodated 64–80 passengers, while the Super B and Series 12 carry 104 and 139 respectively. Maximum cruising speed of the Caravelle 12 is 825km/h and range is 3,465km. The Caravelle is still in use and a total of 279 were built, excluding prototypes. *Country of origin:* France. *Main silhouette:* Caravelle 11R; *top and bottom side views:* Caravelle 6R and 12 respectively. *Picture:* Caravelle 10B.

Power: 2 × **Spey turbofans**   *Span:* (Series 500): 28.5m   *Length:* 32.61m

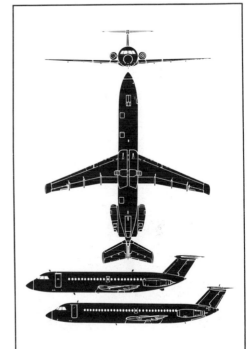

A successful short-haul transport, the One-Eleven first flew in 1963, and remains in production, under licence, in Romania by IAv Bucuresti as the Rombac 1-11. Some 230 One-Elevens were built in the UK and the type has been developed through the Series 200, 300, 400, 475 and 500. The short-fuselage versions carry up to 89 passengers, while the long-fuselage Series 500 can accommodate up to 119. Maximum cruising speed is 871km/h, and range up to 3,000km. The Series 475 is intended for operations from very short airfields. Romanian designations are series 495 and series 560. *Country of origin:* UK. *Main silhouette:* Series 500; *upper side view:* Series 400. *Picture:* Series 560.

Fokker F28 Fellowship

Confusion: One-Eleven, Gulfstream II, Trident, 'Crusty'

Power: 2 × Spey turbofans   *Span:* (Mk 6000): 25.07m   *Length:* (Mk 6000): 26.76m

The Fellowship twin-engined short-haul transport first flew in 1967 and 241 were built, production ceasing in 1987. The different versions are: Mk 1000, short fuselage, up to 65 seats; Mk 2000, longer fuselage, 79 seats; Mk 3000, like 4000 but with short fuselage, 65 passengers; Mk 4000, long fuselage, 85 passengers; Mk 5000, wings of Mk 6000 with 65-seat cabin; Mk 6000, long-span wings and long fuselage with up to 85 seats. Fellowships have also been sold for cargo and VIP use. A distinctive feature of the F28 is the large twin-shell airbrake in the tail. Cruising speed is 670km/h. *Country of origin:* Netherlands. *Silhouette and picture:* Mk 4000.

Power: 2 × Tay turbofans   *Span:* 28.08m   *Length:* 35.53m

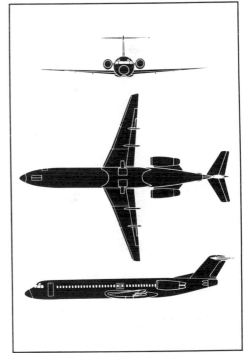

First flown in November 1986, the 100/110-seat Fokker 100 is a complete redesign of the F28 which rivals the BAe 146. Short Brothers are responsible for wing development and MBB for large parts of the fuselage and tail. Range, with 107 passengers, is 2,418km. Deliveries began early in 1989 and, by March 1989, there were firm orders for 212 plus 178 options. *Country of origin:* Holland.

McDonnell Douglas DC-9

Confusion: 'Crusty', One-Eleven, Gulfstream II, Fellowship

Power: 2 × JT8D turbofans   *Span:* (Series 40): **28.47m**   *Length:* (Series 40): **38.28m**

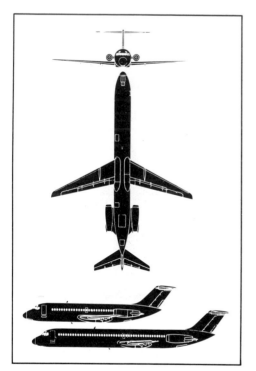

First flown in 1965, the DC-9 was steadily developed and stretched to meet a demand for more seats. Initially, as the DC-9 Series 10, the aircraft carried up to 90 passengers; the long-fuselage Series 30 then appeared, with extended wings and capacity for up to 119. Extended wings and short fuselage resulted in the Series 20, while further fuselage stretches produced the Series 40 (125 passengers) and Series 50 (139 passengers). Military variants are designated C-9A Nightingale and VC-9C by the USAF and C-9B Skytrain II by the USN. Around 1,000 DC-9s were ordered. The Series 40 cruises at 821km/h and has a range of up to 1,723km. *Country of origin:* USA. *Main silhouette:* DC-9-50; *upper side view:* DC-9-10. *Picture:* DC-9-30.

McDonnell Douglas MD-80

Power: 2 × JT8D turbofans   *Span:* 32.85m   *Length:* 41.3m

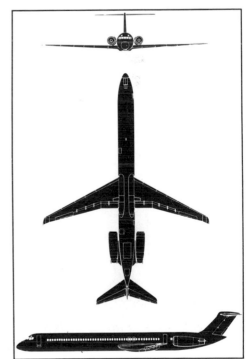

Formerly known as the DC-9 Super 80, the MD-80 is probably the ultimate in stretching. It is 13.26m longer than the original DC-9 Series 10, while passenger capacity has gone up from 90 to 172. Wing area is 28 per cent greater than that of the DC-9 Series 50 and tailplane area is enlarged. Three versions of the MD-80 family are in production, MD-81, MD-82 and MD-83, differing primarily in engines, weights and ranges. The MD-87 is a short fuselage (36.30m) version of the MD-80. The MD-88 has internal and systems refinements. The MD-83 can carry 155 passengers more than 4,670km with a cruising speed of Mach 0.76. The MD-80 first flew in 1979 and by March 1989 the company had received 1,538 orders and options for the MD-80 series. *Country of origin:* USA. *Picture:* MD-83.

Tupolev Tu-22 'Blinder' *Confusion:* 'Fiddler'

Power: 2 × reheated turbojets *Span:* 27.74m *Length:* 40.5m

A most unusual military design, the 'Blinder' bomber has two large turbojets mounted above the rear fuselage, a sharply swept wing, and main undercarriage housed in wing fairings. First flown around 1960, the 'Blinder' is capable of Mach 1.5 at altitude. In basic form, as 'Blinder A', the type was used as a reconnaissance bomber, while the 'B' carried a 'Kitchen' missile under the fuselage. The maritime reconnaissance 'Blinder C' is also equipped with a variety of electronic intelligence equipment. A trainer version, 'Blinder D', has an additional raised cockpit. A flight refuelling probe is carried in the nose. *Country of origin:* USSR. *Silhouette:* 'Blinder C'. *Picture:* 'Blinder B'.

Power: 2 × Soloviev turbofans   *Span:* (Tu-134): 29m   *Length:* 34.35m

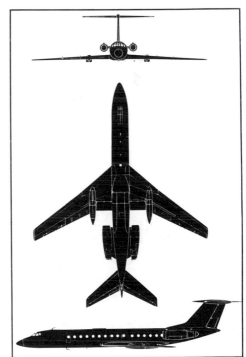

Designed as a successor to the Tu-124 'Cookpot', the Tu-134, NATO code-named 'Crusty', followed the contemporary Western trend of rear-mounted turbofans and T-tails. The Tu-124-type main undercarriage, housed in wing fairings, was retained. First flown in 1962, the 72-seat type went into service with Aeroflot in 1967. A stretched version, the 80-seat Tu-134A, entered service in 1970. Some Tu-134As have the glazed nose replaced with a streamlined radome. The Tu-134 has a maximum cruising speed of 870km/h and a range of 2,400km. Internal modification schemes converted aircraft to B, B-1 and B-3 standard. Over 200 Tu-134/134As were built and a number are in use with Warsaw Pact airlines. *Country of origin:* USSR. *Silhouette and picture:* Tu-134A.

Tupolev Tu-154 'Careless'

Confusion: Trident, 727, DC-9, One-Eleven

Power: 3 × Kuznetsov or Soloviev turbofans   *Span:* 37.55m   *Length:* 47.9m

The Tu-154, code-named 'Careless', was designed as the Soviet counterpart to the Boeing 727 and the Trident, but with the added ability to operate from short, rough runways. The trijet, T-tail 'Careless' retains the characteristic Tupolev main undercarriage housed in wing fairings. First flown in 1968, the 'Careless' entered service with Aeroflot in 1972 and has been exported to other Eastern European countries and to Egypt. The latest version is the Tu-154M with more economical Soloviev engines and certain airframe modifications. The -154M seats up to 180 passengers and has a range of up to 6,600km. *Country of origin:* USSR.

Power: 3 × **Spey turbofans**   *Span:* (2E/3B): **29.87m**   *Length:* (3B): **39.98m**

The Trident was built for the then British European Airways in the early 1960s. The original 1 and 1E versions were followed by the 2E with more range, and the 3B with a longer fuselage. The 3B carries up to 180 passengers compared with 132 on the 2E. The Trident, of which a total of 117 were sold, pioneered the use of Autoland for all-weather operations. China bought a fleet of Trident 2Es and 3Bs. The Trident can cruise at up to 965km/h, and the 2E has a range of 3,965km. *Country of origin:* UK. *Silhouette:* 2E. *Picture:* 3B.

Boeing 727

Confusion: Trident, 'Careless', DC-9, One-Eleven

Power: 3 × JT8D turbofans　　*Span:* (-200): 32.92m　　*Length:* (-200): 46.69m

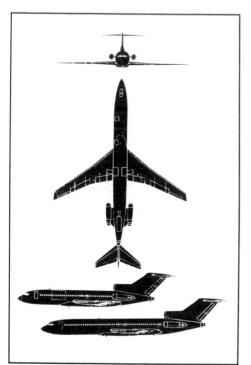

The most widely sold of all Western jet airliners, 1,831 Boeing 727s were built before the type went out of production in August 1984. The only rear-engined aircraft in the Boeing family, the 727 first flew in 1963. The 727-100 seats up to 131 passengers and can be converted to carry cargo. The stretched -200, seating up to 189 passengers, became available in the mid-1960s. This was followed by the Advanced 727-200 with greater all-up weight and more fuel and range. The -200 has range of 4,260km and an economical cruising speed of 917km/h. *Country of origin:* USA. *Main silhouette:* -200; *upper side view:* -100. *Picture:* -200.

Power: 4 × **Conway turbofans**   *Span:* 44.55m   *Length:* (Super VC10): 52.32m

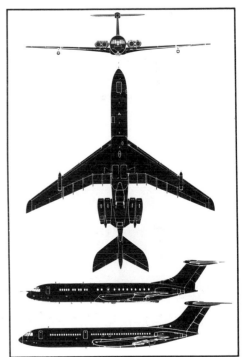

The VC10 was designed to meet a British Overseas Airways Corporation requirement for a long-range airliner with good hot-and-high airfield performance. The type has proved very robust and reliable in service. The VC10 first flew in 1962, followed by the larger, longer-fuselage Super VC10 in 1964. The Super VC10 has a maximum range of 11,470km and cruises at 886km/h. The VC10 is used by the RAF as a standard passenger/freight transport and nine have been converted as three-point in-flight refuelling tankers with more likely to follow. As a tanker, the VC10 is known as the VC10K Mk 2 and the Super VC10 as the VC10K Mk 3. A total of 54 VC10/Super VC10s were built. *Country of origin:* **UK.** *Main silhouette:* **K Mark 2;** *upper side view:* VC10. *Picture:* VC10 C1.

Ilyushin Il-62 'Classic'

Confusion: VC10/Super VC10

Power: 4 × Kuznetsov or 4 × Soloviev turbofans   *Span:* 43.2m   *Length:* 53.12m

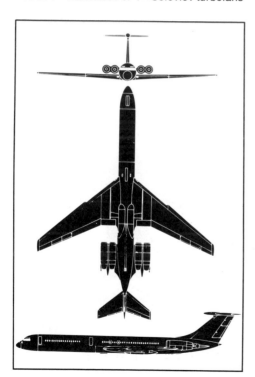

The Soviet Union's first commercial four-engined long-range jet transport, the Il-62, code-named 'Classic', first flew in 1963 and went into service with Aeroflot in 1967. 'Classic' and the VC10 are the only rear-engined four-jet airliners to be built. 'Classic' normally seats 186 passengers. The basic aircraft is powered by Kuznetsov engines, while the Il-62M, introduced in 1974, has Soloviev turbofans. The Il-62M has more power, more fuel and better pay-load range figures. The aircraft cruises at up to 900km/h and in Il-62M form has a maximum range of 10,300km. A high-density variant, the Il-62M-MK, can seat up to 195 passengers. Range with maximum payload 23,000kg and reserves is 8,000km. *Country of origin:* USSR. *Silhouette and picture:* Il-62M.

Power: 3 × Lotarev turbofans   *Span:* 34.90m   *Length:* 36.38m

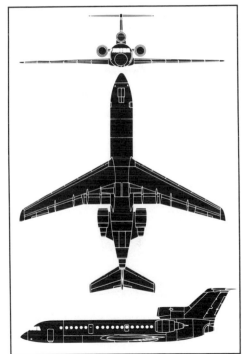

Following on from the straight-wing Yak-40 'Codling', the Yakovlev bureau designed the larger swept-wing Yak-42, code-named 'Clobber', which first flew in March 1975 and went into service with Aeroflot in 1980. The 'Clobber' is replacing the Tu-134 'Crusty' and it is expected that up to 2,000 of the type may ultimately be built. The cabin seats up to 120 passengers and the range of the aircraft is 2,000km, cruising at 750km/h. 'Clobber' is being offered for export in straight passenger form and as a combined passenger/cargo aircraft. A stretched 140 seat version has been built. *Country of origin:* USSR.

BAe 125 *Confusion:* Falcon, Corvette

Power: 2 × Viper turbojets or 2 × TFE731 turbofans   *Span:* 14.33m   *Length:* (Series 700): 15.46m

Some 573 BAe 125 twinjet business aircraft had been sold worldwide before the Series 800 was introduced. Main versions were the Series 400, 600 and 700. The 600 has a longer fuselage than earlier variants and seats up to 14 passengers. All 125s up to and including the Series 600 are powered by Rolls-Royce Viper engines. Thereafter, the company introduced the Series 700, with Garrett AiResearch turbofans and a longer, more pointed nose. The 125-700 cruises at 747km/h and has a range of 4,318km. The Viper-powered 125s used in the RAF are called the Dominie T1 as trainers and CC1 and CC2 as communications aircraft. *Country of origin:* UK. *Silhouette:* Series 700. *Picture:* Dominic TI.

Power: 2 × TFE 731 turbofans   *Span:* 15.66m   *Length:* 15.6m

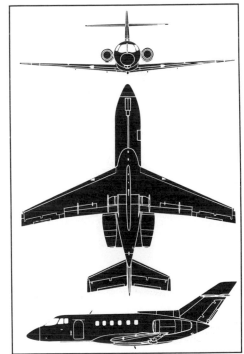

Successor to the 600 and 700 Series 125s, the Series 800 has a redesigned wing, changed fin shape, curved windscreen and the prominent rear underkeel of the earlier versions is deleted. Range, with maximum payload, has been increased to 5,318km. The USAF ordered six 125 series 800 for combat flight inspection and navigation work as the C-29A. BAe and Rockwell are offering the C-29 as a trainer for heavy jet pilots. If the bid is successful, the type will be assembled at Palmdale. By March 1989 a total of 726 125s of all types had been ordered. *Country of origin:* UK.

Sabreliner

Confusion: BAe 125, Falcon, Corvette

Power: 2 × JT12A turbojets or 2 × CF700 or TFE731 turbofans   *Span:* (Series 65): 15.37m   *Length:* 14.30m

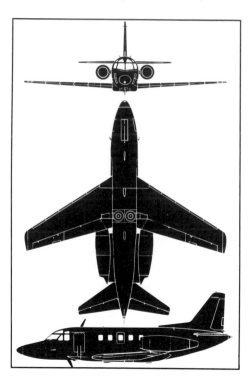

Originally supplied in quantity to the USAF and US Navy as the T-39 and CT-39, the Rockwell Sabreliner was developed into the Series 40 and 60 executive transports. The former carries nine passengers and the latter ten in a lengthened fuselage, both powered by JT12A engines. The 75A is powered by CF700 turbofans and has various refinements, including a wider-span tailplane. Over 600 Sabreliners have been delivered including the Sabreliner 65, with TFE731 engines and a redesigned wing. Eight passengers can be carried. In 1983 the Sabreliner Corp. took over the Sabreliner Division of Rockwell International to provide support for aircraft in service and to provide modifications. *Country of origin:* USA. *Silhouette:* Series 75A. *Picture:* Series 65.

Power: 2 × JT15D turbofans   *Span:* (Beechjet): 13.25m   *Length:* (Beechjet): 13.15m

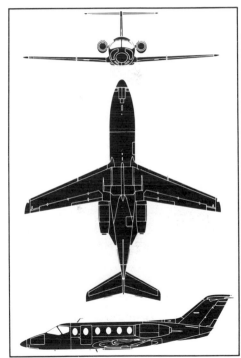

Originally made in component form in Japan, the Mitsubishi Diamond twin-engined executive aircraft was assembled in Texas. The 6–8 passenger Diamond I first flew in August 1978, followed on the production line by the uprated Diamond IA. The Diamond II with improved payload and performance flew in June 1984. In December 1985 Beech Aircraft Corporation acquired the Diamond II programme and, as the Beechjet, it is now manufactured in Salina and Wichita. The Beechjet has a range of 3,575km with a cruising speed of 719km/h. *Country of origin:* USA/Japan. *Picture:* Beechjet.

Dassault-Breguet Mystère-Falcon 10/100

Confusion: Falcon 20, BAe 125, Corvette

Power: 2 × TFE731 turbofans   *Span:* 13.08m   *Length:* 13.85m

Essentially a scaled-down version of the Mystère-Falcon 20, the Mystère-Falcon 10 executive aircraft first flew in December 1970 and deliveries began in 1973. The 10 was replaced on this production line by the 100 with increased take-off weight and a fourth cabin window on the starboard side. The French Navy uses seven 10s as Mystère-Falcon 10 MER for fighter intercept training. Several hundred of the two types have been ordered. Typical weight of the 100 with four passengers, two pilots, and maximum fuel, is 8,280kg. Fast cruise is 912km/h and range 3,480km. *Country of origin:* France. *Silhouette:* Mystère-Falcon 10. *Picture:* Mystère Falcon 100.

Power. 2 × TFE731 turbofans   *Span:* 16.05m   *Length:* 16.94m

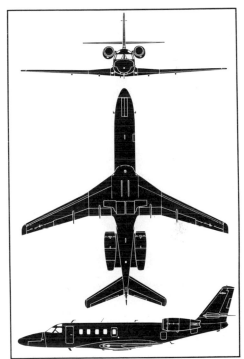

Developed from the straight wing Westwind, the swept-wing 1125 Astra is a business aircraft seating two crew and up to six passengers. Maximum cruising speed is 862km/h and maximum range with long-range tanks and four passengers is 5,763km. The Astra first flew in March 1984. An Israel Aircraft Industries subsidiary company in the USA, called Astra-Jet Corporation, has been concerned with marketing the aircraft. *Country of origin:* Israel.

Dassault-Breguet Mystère-Falcon 20/200

Confusion: Corvette, BAe 125, Falcon 10

Power: 2 × CF700 or 2 × ATF3-6 turbofans   *Span:* 16.30m   *Length:* 17.15m

Dassault and Aérospatiale combined to build the successful Mystère 20/Falcon 20 8/10-seat executive aircraft. First flown in 1963, the aircraft is also used as an air force transport, freighter and trainer. Several hundred have been built and the type is in use in many places round the world. Main versions are the 20F and 200, the latter having Garrett AiResearch ATF3-6 engines in place of GE CF700s. The Mystère-Falcon 200 maritime surveillance aircraft is known in the French Navy as the Gardian and in the US Coast Guard as the HU-25A Guardian. This variant has large side windows for observation. Latest version, for a variety of military purposes, is the Gardian 2. *Country of origin:* France. *Silhouette:* 20F. *Lower side view:* 200. *Picture:* 200.

Power: 3 × TFE731 turbofans   *Span:* 18.86m   *Length:* 18.5m

Seeking more range for executive and other duties, Dassault designed a three-engined development of the Falcon series with new wings but retaining the front and centre-fuselage sections of the Falcon 200. The resulting Falcon 50 seats 8–9 passengers and has a large fuselage fuel tank. Initial deliveries were made in 1979, following a first flight in November 1976. Maximum cruising speed is 870km/h and range with four passengers is 5,560km. The order book totals over 180. *Country of origin:* France.

Dassault-Breguet Mystère-Falcon 900

Confusion: Falcon 100, 200, 50

Power: 3 × TFE-731 turbofans   *Span:* 19.33m   *Length:* 119m

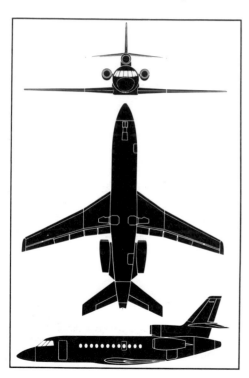

First flown in September 1984, the Mystère-Falcon 900 is an intercontinental tri-jet executive transport seating 19 passengers. Similar in layout to the Mystère-Falcon 50, the 900 has increased dimensions and a larger fuselage. Range, cruising at Mach 0.75, with maximum fuel and eight passengers, is 7,035km. Production Mystère-Falcon 900s commenced delivery in December 1986. Japan has ordered two maritime surveillance equipped Falcon 900s. *Country of origin:* France.

Aérospatiale SN 601 Corvette

Power: 2 × JT15D turbofans   *Span:* 12.87m   *Length:* 13.83m

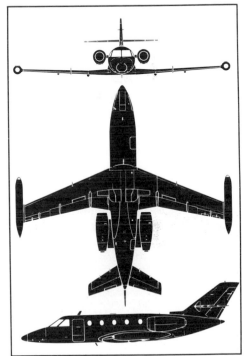

Produced only in small numbers, the Corvette is a 6/14-passenger executive/air taxi/freighter aircraft first flown as the SN 600 in the summer of 1970. Maximum cruising speed is 760km/h and range 1,555km. Fuel tanks are fitted to the wingtips. A crew of two is carried. Production was terminated after completion of the 40th example. Performance figures include an economical cruising speed of 566km/h at 11,900m; maximum rate of climb at sea level of 823m/min; and service ceiling of 12,500m. *Country of origin:* France.

Lockheed JetStar

Confusion: BAe 125, Falcon 10, Falcon 20, Corvette, Challenger

Power: 4 × JT12 or 4 × TFE731 turbofans   *Span:* 16.16m   *Length:* 18.42m

Derived from the original twin-Orpheus JetStar I, the JetStar II is powered by four TFE731 turbofans. The JetStar II seats up to ten and cruises at 817km/h. Differences from earlier versions include a new engine nacelle shape, shortened long-range tanks on the wings, and the addition of a small air intake in the base of the fin. Production is complete. *Country of origin:* USA. *Silhouette and picture:* JetStar II.

Power: 2 × ALF502 or 2 × CF34 turbofans   *Span:* 18.83m   *Length:* 20.82m

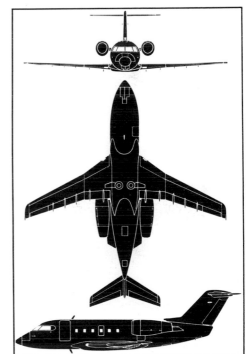

Designed for use as an executive jet, commuter airliner and express cargo aircraft, the Canadair CL-600 Challenger (later re-designated Challenger 600) first flew in August 1978. It has a roomy cabin seating up to 28 passengers and has a cruising speed of 745km/h. The next version is the Challenger 601 with GE CF34 engines and winglets. By March 1989 some 107 Challengers had been ordered. A number of Challengers have been ordered for military use and as VIP transports. A long range development, now available, has a range of 6,667km, while an upgraded engine variant is the 601-3A. A new stretched version is the 601RJ (Regional Jet). *Country of origin:* Canada. *Silhouette:* Challenger 600. *Picture:* Challenger 601.

MBB HFB 320 Hansa

Confusion: Learjet 35/36

Power: 2 × CJ610 turbojets   *Span:* 14.49m   *Length:* 16.61m

Remarkable in having forward-swept wings, the HFB 320 Hansa business jet and light transport first flew in 1964. In executive form the Hansa carries seven/nine passengers and two crew, while up to 12 passengers can be accommodated in a high-density layout. Fifty Hansas were built, including eight for the German Air Force. Production of the type is complete. The Hansa cruises at 825km/h and has a range of 2,370km. Maximum payload of the passenger version is 1,775kg. *Country of origin:* West Germany.

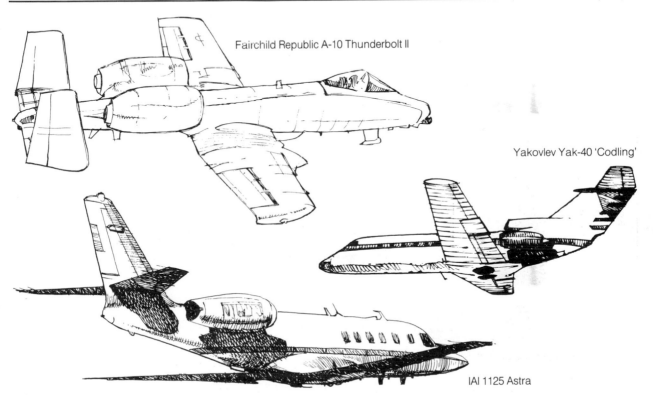

Fairchild Republic A-10 Thunderbolt II

Yakovlev Yak-40 'Codling'

IAI 1125 Astra

Cessna Citation I/II *Confusion:* PD-808, Westwind

Power: 2 × JT15D turbofans   *Span:* (Citation I): 14.35m   *Length:* (Citation II): 14.39m

Initially known as the Fanjet 500, the Cessna Citation seven/eight-seat twin-turbofan executive transport first flew in 1969. The original production version, the Citation 500, was superseded by the Citation I with longer-span wings and uprated turbofans. The Citation II has increased span, a longer fuselage and accommodation for up to 10 passengers. *Country of origin:* USA. *Main silhouette:* Citation 500; *upper side view:* Citation II. *Picture:* Citation II.

Power: 2 × JT15D turbofans   *Span:* 15.9m   *Length:* 14.39m

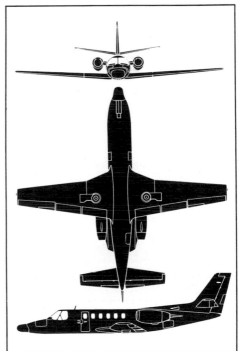

The Citation 11 was replaced on the production line in 1984 by the six/eight seat S/11 version having a revised wing with extended inboard leading edges and modified engine cowlings. The full production S/11 flew in February 1984. Cruising speed is 746km/h. The US Navy ordered 15 Citation S/11s as trainers under the designation T-47A; these have shorter, 14.18m span wings. The Citation V is a recent development with a fuselage 0.6m longer and with an extra window on each side. First customer deliveries were made early in 1989. Many hundreds of Citations, in all variants, have been built. *Country of origin:* USA.

Piaggio PD-808 *Confusion:* Learjet 23/24/25

Power: 2 × Viper turbojets   *Span:* 11.43m   *Length:* 12.85m

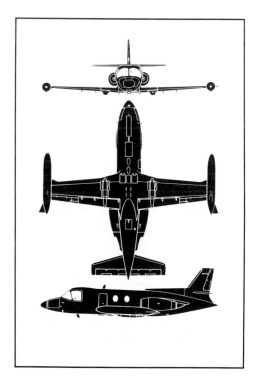

A light jet utility/executive aircraft, the PD-808 was first flown in 1964 and ordered by the Italian Defence Ministry. A maximum of 10 passengers can be seated in the cabin, cruising speed is 722km/h and range is 2,045km. The Italian Air Force uses the PD-808 as a transport, VIP aircraft, navigation trainer and ECM aircraft. All versions carry one or two flight crew; the VIP transport carries six passengers and the electronic countermeasures version is operated by two pilots and three equipment specialists. The designation PD recalls the two companies which launched the project, Piaggio and Douglas. *Country of origin:* Italy.

Power: 2 × CJ610 turbojets   *Span:* 10.84m   *Length:* (25D): 14.5m

Flown for the first time in 1963, the Gates Learjet series of executive transports has been steadily developed and built in large numbers. With six seats and two crew, the Learjet 24 cruises at 774km/h. The tailplane/fin bullet of earlier versions was deleted from the Model 24. The Learjet 25 is 1.27m longer and seats two extra passengers. The latest version is the 25D. A belly camera installation has been fitted to some aircraft. *Country of origin:* USA. *Silhouette:* Learjet 24D. *Picture:* Learjet 24.

Gates Learjet 35/36

Confusion: Learjet 23/24/25

Power: 2 × TFE731 turbofans   *Span:* 12m   *Length:* 14.82m

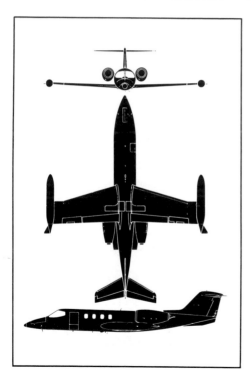

A new series of enlarged Learjet executive aircraft was introduced in 1973. Both wingspan and length were increased, there are five starboard windows instead of four, and the engine nacelles are more bulbous. The wingtips have short rectangular extensions on which the fuel tanks are mounted. Like the Learjet 25, the 35/36 seats eight passengers. Currently in production are the 35A and six-seat, longer-range 36A. A variety of military and paramilitary versions are offered and 80 aircraft are in service with the USAF, designated C-21A. Special mission versions operating in 20 countries are designated EC-35A (electronic warfare and simulation), PC-35A (maritime patrol), RC-35A (reconnaissance), VC-35A (utility version) and U-36A (modified for Japanese Maritime Self Defence Force). *Country of origin:* USA. *Silhouette:* Learjet 35. *Picture:* Learjet 35A (with towed target underneath).

Power: 2 × TFE731 turbofans *Span:* (55): 13.35m *Length:* (55): 14.53m

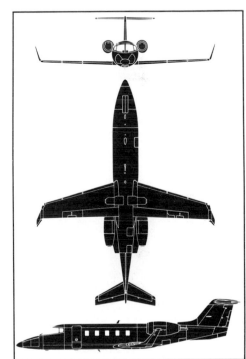

The Gates Learjet 28 Longhorn, an improved version of the Learjet 25, has a redesigned wing with vertical winglets at the tips. The 29 Longhorn is externally similar but with increased fuel capacity. Gates Learjet married a new wide-body ten-passenger fuselage to the wing of the 28/29 to produce the 55 Longhorn. The first 55 Longhorn flew in April 1979 and the first production aircraft in August 1980. Variants include the 55C, 55C/ER with extended range and 55CLR (long range). Extended range versions are the 55ER and LR. Using the fuselage and power plant of the Learjet 35A and 36A combined with similar wings to the Learjet 28/29 the company has produced the Learjet 31 which also has lower rear fuselage strakes. *Country of origin:* USA. *Silhouette:* 55 Longhorn. *Picture:* 28 Longhorn.

IAI Westwind

Confusion: Learjet 23/24/25

Power: 2 × TFE731 turbofans   *Span:* 13.65m   *Length:* 15.93m

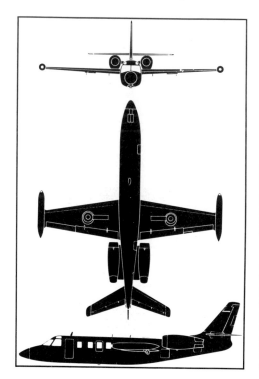

Originally powered by two CJ610 turbojets and known as the IAI 1123, the Westwind executive aircraft (1124) has two Garrett-AiResearch turbofans. Produced by Israel Aircraft Industries, the Westwind carries two pilots and up to ten passengers. Economical cruising speed is 741km/h, and with seven passengers range is over 4,490km. A naval tactical support and coastguard version is available, designated 1124 Sea Scan. A longer-range development is the Westwind 2 with improved wing and winglets on the tip tanks. After the production of 250 Westwinds, the line was closed. *Country of origin:* Israel. *Silhouette:* IAI 1124. *Picture:* Westwind 2.

Yakovlev Yak-40 'Codling'

Power: 3 × Ivchenko turbofans   *Span:* 25m   *Length:* 20.36m

Several hundred examples of the Yak-40, code-named 'Codling', have been built and the type is the most widely used aircraft on Soviet internal short-haul services. First flown in 1966, 'Codling' went into service with Aeroflot in 1968 and since then many have been exported. Seating up to 32 passengers, 'Codling' cruises at 470km/h and has a maximum range of 2,000km. A cargo version with a large door on the port side has been built. Most of the many Yak-40s built so far are in service with Aeroflot, some as air ambulances. Examples have been delivered to Italy, and others are in service in Afghanistan, Bulgaria, Czechoslovakia, France, West Germany, Poland and Yugoslavia. Military users include the Soviet and Polish air forces. *Country of origin:* USSR.

Fairchild Republic A-10 Thunderbolt II

Confusion: Citation

Power: 2 × TF34 turbofans   *Span:* 17.53m   *Length:* 16.26m

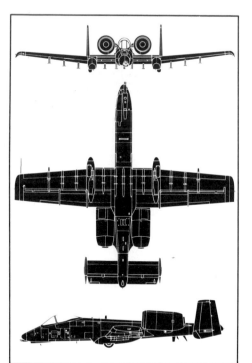

The single-seat A-10A Thunderbolt II is heavily armoured and carries massive armament for ground attack. With a combat speed of 712km/h and a radius of action of 1,000km, the Thunderbolt can carry 5,450kg of weapons externally. In addition to bombs, rockets and guided missiles, the Thunderbolt II has a seven-barrel 30mm gun mounted in the nose. A total of 825 Thunderbolts have been built and the type is in service in the USA and Europe. A combat-readiness trainer version is designated A-10B. *Country of origin:* USA. *Silhouette and picture:* A-10A.

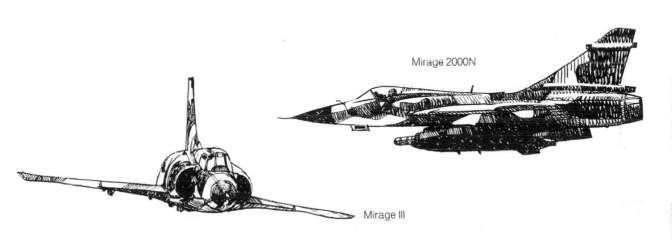

Mirage 2000N

Mirage III

Sukhoi Su-15 'Flagon'

Dassault-Breguet Mirage III

Confusion: Mirage 5, Mirage IV, Mirage 2000, Kfir

Power: 1 × Atar reheated turbojet   *Span:* 8.22m   *Length:* (IIIE): 15.03m

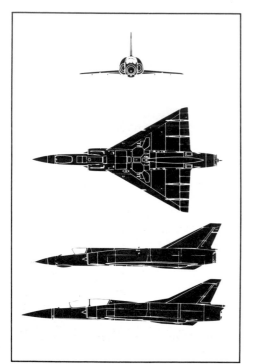

One of the most successful single-seat fighter/ground-attack aircraft produced in Europe, the Mirage III has been sold to many countries. First flown in 1956, it has been built in several versions: IIIC interceptor, IIIE ground attack, IIIB and IIID two-seat trainers and IIIR reconnaissance aircraft. Basic armament consists of two 30mm cannon in the fuselage. Bombs, rockets, guided missiles and long-range tanks can be carried on centreline and underwing pylons. The aircraft is capable of Mach 2.2 and radius of action for ground attack is 1,200km. *Country of origin:* France. *Main silhouette:* Mirage IIIE; *lower side view:* Mirage IIID. *Picture:* Mirage IIIS.

Dassault-Breguet Mirage 5/50

Power: 1 × Atar 9C turbojet   *Span:* 8.22m   *Length:* 15.55m

The ground-attack version of the Mirage III is designated Mirage 5. Over 1,400 III/5s were built. The Mirage 5 has simplified avionics and a longer nose than the Mirage III. The prototype was first flown in May 1967. Armament combinations include cannon, bombs and missiles. Maximum speed of the Mirage 5 is 2,350km/h at 12,000m. Service ceiling is 17,000m. Mirage 5 designations indicate the customer country: 5SDE (*Arabie SaouDitE*—Saudi Arabia) and 5AD (Abu Dhabi), for example. The two-seat trainer has the letter ''D'' after the country designation. A new development is the Mirage 50 with up-rated Atar 9K50 engine and new electronics systems. A variant on offer is the Mirage 3NG with canard surfaces just aft of the intakes, advanced radar and fly-by-wire controls. *Country of origin:* France. *Silhouette:* Mirage 5. *Picture:* Mirage 50.

Dassault-Breguet Mirage 2000

Confusion: Mirage III/5, Mirage IV, Kfir

Power: 1 × M53 reheated turbofan   *Span:* 9m   *Length:* 15.33m

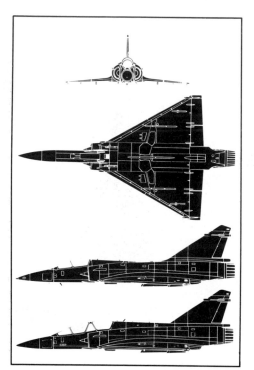

Although bearing a superficial resemblance to the Mirage III/5, the Mirage 2000 is a completely new aircraft which provides the French Air Force with a very advanced Mach 2+ interception capability. The controls are electrically signalled 'fly-by-wire', with computer control. In its primary role the single-seat Mirage 2000 will carry Super Matra 530 and R550 Magic air-to-air missiles and external fuel tanks. A variety of weapons can be carried for ground attack. A two seat trainer variant is designated 2000B. A two seat 'Penetration' version known as Mirage 2000N, with a stand-off nuclear missile, has also been ordered by the FAF. *Country of origin:* France. *Main silhouette:* Mirage 2000; *lower side view:* Mirage 2000N. *Picture:* Mirage 2000.

Dassault-Breguet Mirage IVA/IVP

Power: 2 × Atar reheated turbojets   *Span:* 11.85m   *Length:* 23.5m

The Mirage IVA was designed in the 1950s as France's nuclear deterrent bomber. Its shape owes much to the Mirage III but it is much larger. Seating two in tandem, the Mirage IVA first flew in 1959 and 62 had been completed when production ended in March 1968. There are 30 Mirage IVs still in use. Of these, 18 have been modified as the Mirage IVP. The IVP carries the ASMP tactical nuclear stand-off missile in place of the free-fall nuclear bomb recessed in the fuselage underside. *Country of origin:* France. *Silhouette:* Mirage IVA. *Picture:* Mirage IVP.

IAI Kfir

Confusion: Mirage III, Mirage 5, Mirage 2000, Viggen

Power: 1 × J79 reheated turbojet   *Span:* 8.22m   *Length:* 16.35m

An extensively redesigned Mirage 5 built in Israel and powered by a General Electric J79 engine, the Kfir first went into service with the Israeli Air Force in 1975 as the Kfir-CI. A modified version, the C2, has additional small winglets, or canards, just behind the air intakes and an extended wing leading edge. A dual-role interceptor and ground-attack aircraft, the Kfir has two fuselage-mounted 30mm cannon and seven hard points for air-to-air and air-to-ground missiles, bombs or drop tanks. The Kfir has a top speed of over 2,335km/h. A two-seat trainer derivative is designated Kfir-TC2. Last production version was the Kfir-C7 with improved performance, the trainer version being the TC7. Altogether 212 Kfirs were built. *Country of origin:* Israel. *Main silhouette:* Kfir-C2; *lower side view:* Kfir-TC2. *Picture:* Kfir-C2.

Power: 1 × **Atar reheated turbojet**   *Span:* 8.22m   *Length:* overall: 16.36m

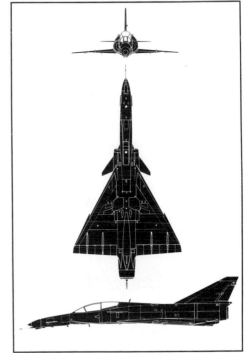

An advanced development of the Mirage 3, the South African Cheetah bears a close resemblance to the Israeli Kfir. Rebuilt from existing Mirage airframes, the Cheetah has a dog tooth leading edge to the wing, canard surface extending from the intake cowling, small strakes on the nose and a revised nose shape. An unusual feature is the weapons pylons mounted at the wingroot/air intake duct junction. The Cheetah is built in single (E2) and two seat (D2) variants and carries two 30mm cannon, air-to-air and air-to-surface missiles and bombs. The internal systems have been completely changed and an in-flight refuelling probe is fitted. *Country of origin:* South Africa. *Picture:* Cheetah E2.

Saab Viggen   *Confusion:* Kfir

Power: 1 × RM8A reheated turbofan   *Span:* 10.6m   *Length:* (including probe): 16.3m

A high-performance Mach 2 multi-role aircraft, the Saab Viggen entered service with the Swedish Air Force in 1971. The Viggen is unusual in having a double-delta layout with the tailplane forward of the wing. The various versions are as follows: AJ37 all-weather attack; JA37 interceptor; SF37 reconnaissance; SH37 maritime reconnaissance; and SK37 two-seat operational trainer with raised rear cockpit. A wide variety of weapons can be carried on three underfuselage and four underwing pylons. The Viggen can operate from unprepared strips and roads. *Country of origin:* Sweden. *Silhouette:* JA37. *Picture:* AJ37.

Aérospatiale/BAe Concorde

Power: 4 × Olympus 593 reheated turbojets   *Span:* 25.56m   *Length:* 61.66m

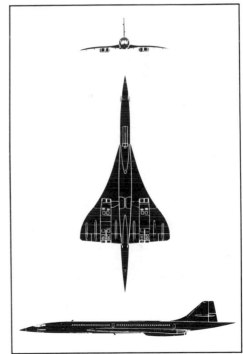

The world's only fully operational supersonic airliner, the Anglo-French Concorde can carry 128 passengers at a speed of Mach 2 for 6,300km. It first went into service with British Airways and Air France in 1976. Eighteen Concordes were built, of which 14 entered service. The external layout is unusual: the engines are boxed under the wing in two pairs, while the wing is ogive-shaped in planform and steeply cambered at the leading edge. Concorde's service ceiling is 18,300m, higher than that of any other airliner in service. Concorde commercial services were inaugurated on January 21, 1976, when the fifth and sixth production aircraft (Air France F-BVFA and British Airways G-BOAA) flew Paris-Dakar-Rio de Janeiro and London-Bahrain respectively. *Countries of origin:* France/UK.

Saab Draken   *Confusion:* F-106

Power: 1 × RM6 reheated turbojet   *Span:* 9.4m   *Length:* 15.4m

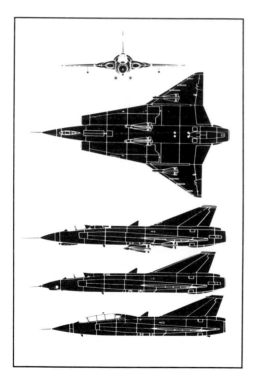

With its unusual double-delta wing and Mach 2 performance, the Saab Draken fighter first flew in 1955. It remained in production until 1976 and is in service with Sweden, Denmark, Austria and Finland. Apart from its high performance, the Draken can operate from small airfields. Various versions have been produced, including the J35A, B, D and F, plus the SK35C two-seat trainer and S35E reconnaissance aircraft. *Country of origin:* Sweden. *Main silhouette:* J35F; *middle side view:* S35E; *bottom side view:* SK35C. *Picture:* J35F.

McDonnell Douglas A-4 Skyhawk

Power: 1 × J65 or J52 turbojet   *Span:* 8.38m   *Length:* 12.27m

Designed as a carrierborne attack aircraft, the A-4 Skyhawk was nicknamed the 'Bantam Bomber'. Although small, it can carry a very heavy weapon load plus two cannon. Variants have gone from A-4A through to -4S. The two-seat trainer is designated TA-4. All versions from the F model onwards carry a dorsal hump containing avionics equipment, while the -4M was the first sub-type to be known as Skyhawk II. The Skyhawk has a top speed of 1,078km/h and a range of 1,480km. Some 2,960 were built and production has been completed. Various up-date programmes are being carried out on A-4s. *Country of origin:* USA. *Main silhouette:* A-4M; *middle side view:* A-4E; *bottom side view:* TA-4F. *Picture:* A-4M.

General Dynamics F-16 Fighting Falcon

Confusion: Skyhawk

Power: 1 × F100 reheated turbojet   *Span:* 9.45m   *Length:* 14.52m

Highly manoeuvrable and capable of speeds in excess of Mach 2, the F-16 fighter is being built in the USA and in NATO countries and is in service with many air forces. A particular feature is the 'shark's mouth' intake. A multi-barrel 20mm cannon is mounted in the fuselage, together with weapons or tanks on four underwing pylons and air-to-air missiles on the wingtips. The production single-seater is known as the F-16A while the tandem two-seat trainer is designated F-16B. A larger tailplane was fitted from late 1981, and the F-16C and D with updated systems are in production. The US Navy uses the F-16N as a supersonic adversary aircraft, while the A-16 is a proposed close-air-support version. *Country of origin:* USA. *Silhouette:* F-16A. *Picture:* F-16D.

Power: 2 × AL-21F reheated turbojets   *Span:* (Flagon-A): 9.15m   *Length:* (including probe): 20.25m

A Mach 2.5 twin-engined all-weather fighter, the Su-15, code-named 'Flagon', is used in large numbers by the Soviet Air Force. In its early form, 'Flagon A', the aircraft had the same wing shape as the 'Fishpot'. The latest variants, 'Flagon D', 'E' and 'F', have wing leading edges with distinctive compound sweep. The 'Flagon' has a massive nose-mounted radome and large air-to-air missiles are carried on the outboard sections of the wing. Like 'Fishpot', 'Flagon' can carry twin auxiliary tanks under the fuselage. The two-seat version is known as 'Flagon C'. 'Flagon F' is distinguished from the 'D' and 'E' by its ogival, not conical, radome. 'Flagon' can carry a belly gun pack. 'Flagon G' is a two seat trainer version of 'Flagon F'. *Country of origin:* USSR. *Main silhouette:* 'Flagon F'; *lower side view:* 'Flagon C'. *Picture:* 'Flagon F'.

SAC J-8 II 'Finback'

Confusion: 'Flagon', 'Fishbed'

Power: 2 × WP-13A reheated turbojets   *Span:* 9.34m   *Length:* 21.59m

A single-seat twin-engined air superiority fighter, the J-8 II, NATO code-named 'Finback', has a design history stretching back to the 1960s. The avionics of the J-8 II are being upgraded by Grumman in the USA, with Westinghouse and Litton as sub-contractors. Export variants may carry alternative avionics. Capable of a maximum level speed of 1300km/h and a combat radius of 800km, the J-8 II has a 23mm twin barrel cannon under the fuselage and a variety of underwing stores, including PL-2B and PL-7 air-air missiles. *Country of origin:* China.

Power: 1 × R-13 reheated turbojet   *Span:* 7.15m   *Length:* (including probe): 15.76m

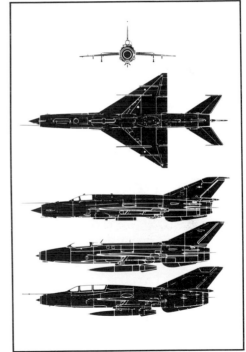

The MiG-21 'Fishbed' fighter is one of the world's most widely used combat aircraft, serving with 28 air forces. 'Fishbed' has been produced in a variety of versions, including the original with a small nose radome 'Fishbed C', all-weather variant with enlarged nose radome 'Fishbed D', main production standard with wider-chord fin 'Fishbed F', deeper dorsal fairing 'Fishbed J', reconnaissance version with pod 'Fishbed H', and latest variants with enlarged dorsal fairing 'Fishbed K', 'L' and 'N'. The two-seater trainer is known as 'Mongol'. Versions of the Mig-21 are in production in China as the Xian J-7/F-7. *Country of origin:* USSR. *Main silhouette:* 'Fishbed J'; *middle side view:* 'Fishbed C'; *lowest side view:* 'Mongol'. *Picture:* 'F-7M'.

Lockheed SR-71A Blackbird   *Confusion: —*

Power: 2 × JT11D reheated turbojets   *Span:* 16.95m   *Length:* 32.74m

The layout of the SR-71A Blackbird high-altitude strategic reconnaissance aircraft is unique, with the engines mounted on the delta wing and twin canted fins on the engine nacelles. Carrying a crew of two in tandem, the aircraft can fly at up to Mach 3 and has an operational ceiling of over 24,384m. A variety of cameras and electronic sensors are carried. Trainer versions are designated SR-71B and -71C. First aircraft in the series was the Falcon missile-armed YF-12A interceptor, which was cancelled. First flight of the SR-71 was in December 1964. Three YF-12As and about 30 SR-17s were built and eight of the latter are believed to be still in service. *Country of origin:* USA. *Silhouette:* SR-71A. *Picture:* SR-71B.

Tornado F2

Su-24 'Fencer'

Tu-26 'Backfire B'

205

Panavia Tornado IDS

Confusion: Tornado ADV, 'Flogger', F-111, 'Fencer'

Power: 2 × RB.199 reheated turbofans   *Span:* (wings fully spread): 13.9m   *Length:* 16.7m

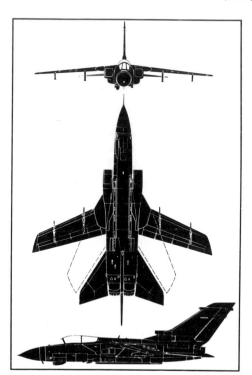

In full scale service, the two-seat Tornado IDS (Interdictor Strike) forms the offensive backbone of the RAF, the West German Air Force and Navy, the Italian Air Force and the Royal Saudi Air Force. Total production of the IDS version is scheduled to reach over 1,000, and in the RAF the type is designated Tornado GR1. Capable of twice the speed of sound at altitude, the IDS can fly at up to 1,480km/h at low level. The IDS can carry a wide range of weapons and two 27mm cannon are fitted in the lower forward fuselage. *Countries of origin:* Germany/Italy/UK.

Power: 2 × RB.199 reheated turbofans   *Span:* (wings fully spread): 13.9m   *Length:* 18.06m

Developed by British Aerospace for the RAF, the Tornado ADV (Air Defence Variant) is the RAF's standard all-weather fighter. With a longer fuselage and Foxhunter airborne interception radar, the Tornado ADV carries a single 27mm cannon and four Sky Flash air-to-air missiles under the fuselage. Two Sidewinder missiles and long-range fuel tanks can be carried under the wings. A Mach 2 aircraft with long endurance, the Tornado ADV is known in the RAF as the F2 and the type has been bought by Saudi Arabia and Oman. Like the Tornado IDS, the ADV is equipped for in-flight refuelling. *Countries of origin:* Germany/Italy/UK.

Grumman F-14 Tomcat

Confusion: F-111, 'Fencer', Tornado

Power: 2 × TF30 or F110 reheated turbofans   *Span:* (wings fully spread): 19.55m   *Length:* 18.9m

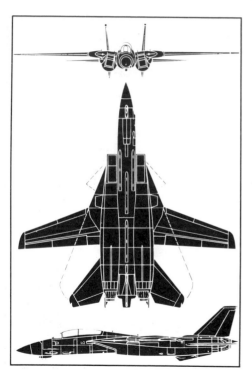

The United States Navy's counterpart to the USAF F-15, the two-seat F-14A Tomcat is an all-weather carrierborne interceptor. With its powerful fire-control system, Tomcat can engage several targets simultaneously with Phoenix, Sparrow or Sidewinder air-to-air missiles. Tomcat is unusual in having twin fins and variable-geometry wings. Performance includes speeds of just over Mach 1 at sea level and Mach 2+ at high altitude. A new version of the Tomcat with General Electric F110 engines is designated F-14D. *Country of origin:* USA.

Power: 2 × TF30 reheated turbofans   *Span:* (wings fully spread): 19.2m   *Length:* 22.4m

This all-weather two-seat strike/attack aircraft, used by the USAF in Europe and America, carries heavy underwing armament and is capable of over Mach 2 at altitude. Versions are the F-111A (141 built), F-111E (94 built), F-111F (82 built) and FB-111A (77 built). All versions have generally similar outlines. Principal performance figures include maximum speeds of 2,655km/h at 15,000m and 1,472km/h at sea level; ceiling of 18,300m; and maximum weapon loads of 24 × 1,000lb or 50 × 750lb bombs. *Country of origin:* USA. *Main silhouette:* FB-111A; *Upper side view:* F-111E. *Picture:* EF-111A.

Mikoyan MiG-23S 'Flogger B'/'E'/'G'/'K'

Confusion: 'Flogger D', Tornado, 'Fencer', F-111

Power: 1 × Tumansky reheated turbofan   *Span:* (wings fully spread): 14.25m   *Length:* 16.8m

A standard Soviet interceptor, this Mach 2+ type is in widespread use in Eastern Europe and has been exported to the Middle East, Far East, Africa and Cuba. It carries cannon and missile armament. A two-seat variant is known as the 'Flogger C', and 'Flogger G' is an improved interceptor. The 'Flogger E' export variant is generally similar to the Soviet Air Force version but is equipped to a lower standard. It is fitted with a smaller radar (NATO code-named 'Blue Jay') in a shorter nose radome, and lacks the undernose laser rangefinder and Doppler navigation equipment of 'Flogger B'. 'Flogger K' has wing aerodynamic modifications, smaller ventral fin and smaller 'Flogger G' type dorsal fin. *Country of origin:* USSR. *Silhouette and picture:* 'Flogger B'.

Power: 1 × Tumansky reheated turbofan   *Span:* (wings fully spread): 14.25m   *Length:* 16.46m

A tactical strike development of 'Flogger B', the 'Flogger D' is one of the major types facing the West in Europe. It differs from 'Flogger B' in having fixed air intakes and tailpipe nozzle, and a laser rangefinder nose. There are five external racks for missiles and bombs. Compared with the MiG-23 'Flogger B', the MiG-27 has a completely redesigned forward fuselage. The ogival radome is replaced with a sharply tapered nose. A Gatling-type gun replaces the interceptor's twin-barrelled weapon. 'Flogger F' is an export version of the MiG-23 with outline as for MiG-27 'Flogger D', while 'Flogger H' is similar but with the addition of avionics pods forward of the nosewheel doors. MiG-27 'Flogger J' has a lipped top to the nose and a nose under-blister. Some 'J's have wing root extensions. *Country of origin:* USSR. *Silhouette and picture:* 'Flogger D'.

Sukhoi Su-17 'Fitter C'   *Confusion:* 'Fitter A', 'Fitter B'

Power: 1 × Lyulka or 1 × Tumansky reheated turbojet   *Span:* (wings fully spread): 13.7m   *Length:* 18.75m

The single-seat 'Fitter C' tactical strike fighter is unusual in being a variable-geometry development of a fixed-wing aircraft. Compared with 'Fitter A', the outer wings pivot, the engine is more powerful, and armament and fuel have been increased. Export versions are designated Su-20 and Su-22. Certain other 'Fitters', such as the 'J', have a changed fin and bulged rear fuselage associated with the fitting of a Tumansky engine. 'Fitter K' has an extended dorsal fin with a cooling intake in front. Operators include the Warsaw Pact countries, Syria and Peru. *Country of origin:* USSR. *Silhouette:* 'Fitter C'. *Picture:* 'Fitter K'

Power: 2 × Tumansky reheated turbofans   *Span:* (wings fully spread): 17.50m   *Length:* 21.29m

One of the advanced types in the Soviet inventory, the Sukhoi Su-24, code-named 'Fencer', is a low-level supersonic attack aircraft in the same class as is the American F-111. Some 500 two-seat 'Fencers' are in service, with production continuing at a high rate. Armament comprises two 30mm cannon in the lower fuselage and nuclear/conventional bombs, rocket projectiles, or missiles on eight wing and fuselage pylons. With wings fully swept to 70°, maximum speed at altitude is in excess of Mach 2. Five versions 'A', 'B', 'C', 'D' and 'E' have been identified. The 'D' has large wing fences integral with pylons, extended fin base and in-flight refuelling. 'Fencer E' is equipped for electronic warfare. *Country of origin:* USSR. *Silhouette:* 'Fencer C'. *Picture:* 'Fencer E'.

Rockwell International B-1B   *Confusion:* 'Backfire'

Power: 4 × F101 reheated turbofans   *Span:* (wings fully spread): 41.66m   *Length:* 44.7m

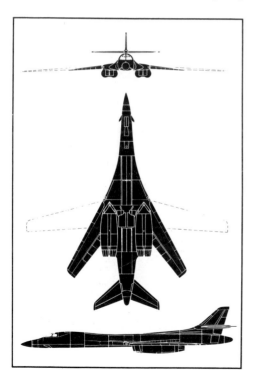

First flown in 1974, the B-1 supersonic long-range bomber was to have been the subject of large orders for the US Air Force. These were cancelled by the Carter administration and only four prototypes were completed. President Reagan revived the project and a new programme for 100 B-1Bs was announced in 1981. Externally similar to the B-1, the B-1B carries a crew of four and is armed with Air Launched Cruise Missiles (ALCMs) and Short Range Attack Missiles (SRAMs). The B-1B can operate low down at supersonic speeds or can achieve speeds of up to Mach 2 at altitude. Maximum range is 9,820km. *Country of origin:* USA.

Power: 2 × Kuznetsov reheated turbofans　　*Span:* (wings fully spread): 34.30m　　*Length:* 39.6m

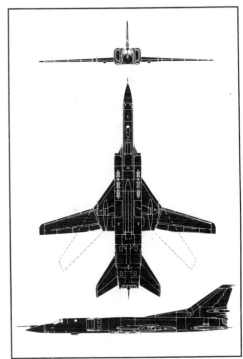

The Soviet strategic bomber force consists largely of 'Backfires'. Carrying either nuclear or conventional weapons, 'Backfire' can fly a combat radius of 4,000km and is capable of supersonic speeds at all altitudes. Three versions exist: the preliminary 'Backfire A', the definitive 'B', and 'Backfire C' with wedge type intakes. Production of Backfire is running at a rate of about 30 a year, and over 350 are in service. 'Backfire' will operate alongside the much larger 'Blackjack' detailed on the next page. *Country of origin:* USSR. *Silhouette:* 'Backfire B'. *Picture:* 'Backfire C'.

Tupolev Tu-160 'Blackjack'

Confusion: 'Backfire', B-1

Power: 4 × reheated turbofans   *Span:* (wings fully spread): 55.7m   *Length:* 54m

The Tu-160 'Blackjack' variable geometry wing strategic bomber is a completely new design and not just a development of 'Backfire'. It is believed to have entered service in 1988 and is in full scale production. Maximum speed is Mach 2 and unrefuelled radius of action approximately 7,300km. 'Blackjack's' armament consists of air-launched cruise missiles and bombs. No gun armament is carried. *Country of origin:* USSR.

Northrop B-2

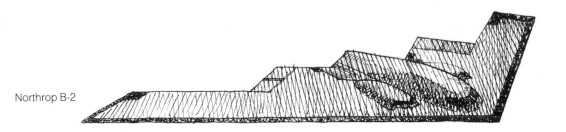

Lockheed F-117A

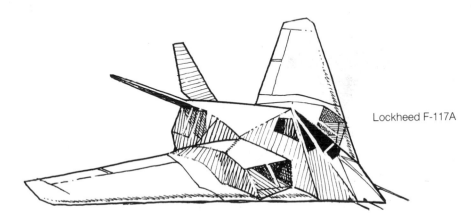

Lockheed F-117A

Confusion: —

Power assumed: 2 × F404 turbofans   *Span:* (approx): 13.7m   *Length:* (approx): less than 10.66mm

The subject of world speculation for a number of years, the single seat F-117A 'Stealth' fighter made its first flight in June 1981 and became operational in 1983. Resembling an arrowhead in planform, the aircraft has been designed to have numerous facets which make it difficult to be seen by radar. In addition it has low noise characteristics and low infra red and smoke emissions from the engines. Basically the F-117A is a squat, short, angular central body mounted on swept wings and with a sharply canted vee tail. At the time of writing no official dimension or performance figures had been revealed. *Country of origin:* USA.

Power: 4 × F118 turbofans   *Span:* 52.4m   *Length:* 21m

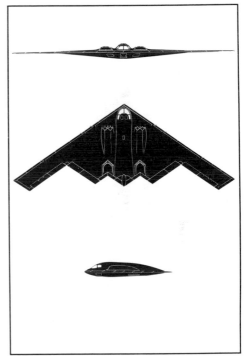

The second 'Stealth' aircraft to be revealed is the Northrop B-2 bomber. The company has adopted a different approach to the F117A in trying to reduce radar reflection, infra red imprint, etc. The B-2 is that rare specimen, a true flying wing without fins. The swept wing itself has a 'sawtooth' trailing edge and the twin intakes and exhausts for the four engines are in the wing upper surface. The B-2 has three crew and can carry 16 SRAM missiles or nuclear bombs. The programme involves a prototype and 132 production aircraft, the first base to receive B-2s being Whiteman AFB, Missouri in 1991. The B-2 first flew in July 1989. *Country of origin:* USA.

Jet, straight wing, fuselage/wing engine(s)

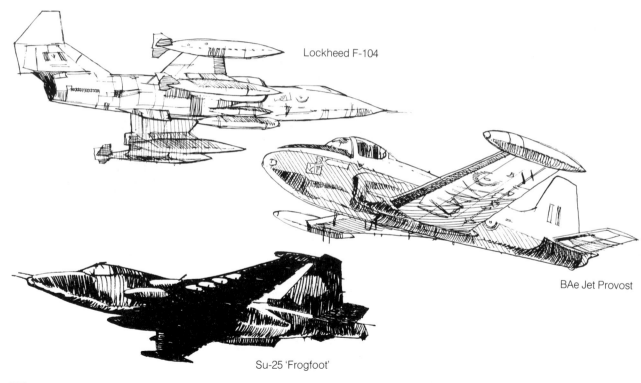

Lockheed F-104

BAe Jet Provost

Su-25 'Frogfoot'

Canadair CL-41 Tutor

Power: 1 × J85 turbojet   *Span:* 11.13m   *Length:* 9.75m

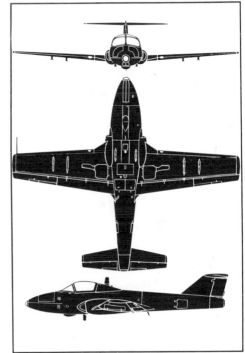

Originally developed as a private venture, the Canadair CL-41 Tutor side-by-side two-seat trainer was adopted by the Royal Canadian Air Force. Production was completed in 1966 after 190 had been built. Twenty CL-41Gs with fuselage pylons were built for Malaysia, named Tebuan. These aircraft are believed to be in store. Basic armament comprises bombs, rockets and gun pods on wing hardpoints. Maximum speed is 774km/h. *Country of origin:* Canada. *Silhouette and picture:* CL-41G.

Morane-Saulnier MS-760 Paris

Confusion: Delfin, Tutor

Power: 2 × Marboré turbojets   *Span:* 10.15m   *Length:* 10.05m

A four-seat high-speed civil and military communications aircraft, the MS-760 Paris first flew in July 1954. Some 165 Paris Mks I and II were built. The Paris II has higher-powered engines and more fuel capacity. A prototype of the Paris III, with six seats and extended wingtips, was built. The Paris I has a maximum speed of 650km/h and a range of 1,500km. For weapon training the Paris II can carry 7.5mm machine guns and guns and bombs or rockets. The Paris remains in service in France and Argentina. *Country of origin:* France. *Silhouette and picture:* Paris I.

Power: 1 × M-701 turbojet   *Span:* 10.29m   *Length:* 10.81m

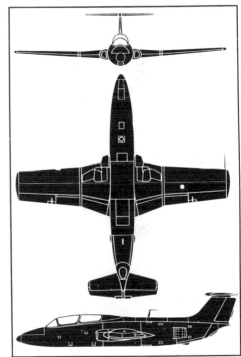

Designed as a Communist-bloc basic trainer, the L-29 Delfin, NATO code-name 'Maya', first entered service in 1963. The air forces of Czechoslovakia, the Soviet Union, East Germany, Bulgaria, Romania and Hungary use the Delfin, and other countries outside the Warsaw Pact have been supplied with the type. Two underwing pods can carry bombs, rockets, drop tanks or 7.62mm guns. An aerobatic version of the Delfin with a single seat is designated the L-29A Akrobat. *Country of origin:* Czechoslovakia. *Silhouette and picture:* L-29.

223

RFB Fantrainer   *Confusion:* Delfin, CL-41

Power: 1 × Allison 250 turboshaft   *Span:* 9.7m   *Length:* 9.25m

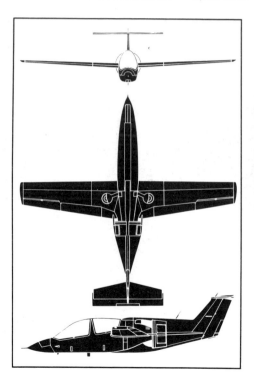

Although not strictly a 'jet', the Fantrainer looks exactly like one in the air, albeit without a rear jetpipe. A unique two-seat basic trainer, the Fantrainer has a turboshaft engine driving a ducted fan in the fuselage just aft of the wings. First flown in 1978, the aircraft is on offer in two forms, the model 400 with a 420hp engine and the 600 with a 650hp unit. The Royal Thai Air Force has bought 47 Fantrainers and two are being leased to Lufthansa. Take-off weight of the model 400 is 1,799.7kg and maximum speed 370km/h. *Country of origin:* West Germany. *Picture:* Fantrainer model 600.

Confusion: Strikemaster   # Cessna A-37 Dragonfly

Power: 2 × J85 turbojets   *Span:* 10.3m   *Length:* 8.93m

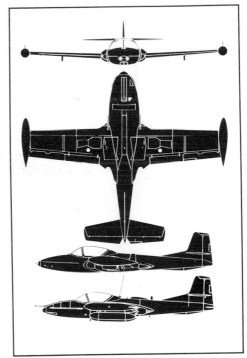

The standard USAF basic trainer (T-37B), the Dragonfly originally flew with Continental T69 engines. Many Dragonflies are in service with a number of air forces as trainers, with underwing pylons (T-37C) and with a Minigun in the fuselage and four weapon stations on the wings (A-37). As a forward air control aircraft the Dragonfly is designated OA-37B. A large number of Dragonflies have been built. Maximum speed is 816km/h and range with weapon load is 740km. *Country of origin:* USA. *Main silhouette:* T-37A; *lower side view:* A-37C. *Picture:* A-37B.

Lockheed T-33A Shooting Star

Confusion: Galeb

Power: 1 × J33 or 1 × Nene turbojet   *Span:* 11.85m   *Length:* 11.48m

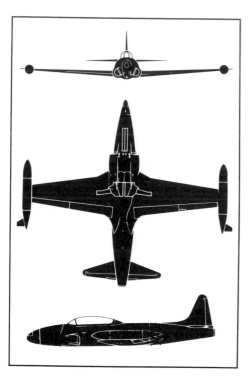

Trainer version of the US Air Force's first operational jet fighter, the T-33A (originally TF-80C) first flew as far back as 1948. A total of 6,600 T-33s were built and the type is still in service in some numbers. Some Shooting Stars were built with noses modified for reconnaissance. Several air forces still operate T-33s, including the Canadian Forces where the type is known as the CT-133. Maximum speed is 960km/h and endurance 3.12h. *Country of origin:* USA. *Picture:* CT-133.

Power: 1 × SO-1 turbojet   *Span:* 18.18m   *Length:* 12.74m

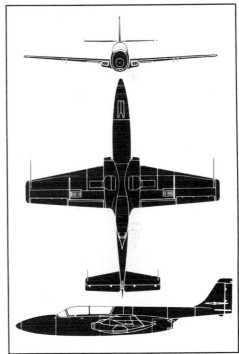

The Polish Air Force took delivery of its first TS-11 Iskra tandem two-seat aerobatic trainer in 1963. The layout is unusual, with the engine exhausting under the fuselage and the tail assembly being carried by a long boom. Forward-firing machine guns are housed in the fuselage, and weapons can be carried on four underwing pylons. Several hundred Iskras have been built. Main performance figures include a maximum level speed at 5,000m of 720km/h; maximum rate of climb at sea level of 888m/min and service ceiling of 11,000m. The Iskra is operated by Poland and India. *Country of origin:* Poland.

SOKO G-2 Galeb/Jastreb *Confusion:* Iskra, Delfin

Power: 1 × Viper turbojet   *Span:* 10.47m   *Length:* 10.34m

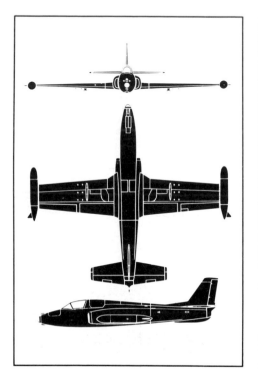

The first Yugoslav jet to enter production, the G-2 Galeb was first flown in 1961 and is now the standard Yugoslav Air Force basic trainer. Also in service is the TJ-1 Jastreb single-seat ground-attack version. This is equipped with three nose machine guns and eight underwing pylons for weapons, compared with the Galeb's two machine guns and smaller weapon load. Maximum speed is 756km/h and maximum endurance 2hr 30min. *Country of origin:* Yugoslavia. *Silhouette and picture:* Jastreb.

Power: 1 × Viper turbojet   *Span:* 10.7m   *Length:* 10.6m

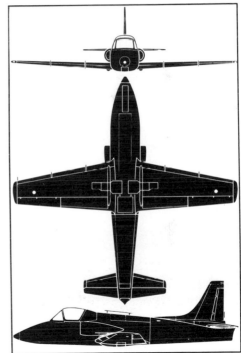

Designed to replace the Vampire trainers licence-built in India, the HJT-16 Kiran basic trainer was first flown in 1964. The aircraft bears a superficial resemblance to the Jet Provost/Strikemaster but is in fact totally of Indian design. The crew of two are seated side-by-side. A weapon-carrying variant is the Mk IA. A development of the Kiran I is the Orpheus-powered Mk II, intended for counter-insurgency work and weapons training. Two fuselage machine guns and four underwing pylons are fitted. Maximum speed of the Kiran I is 659km/h. *Country of origin:* India. *Silhouette:* Kiran I. *Picture:* Kiran II.

BAe Strikemaster

Confusion: Kiran, MB-326

Power: 1 × Viper turbojet   *Span:* 10.77m   *Length:* 10.27m

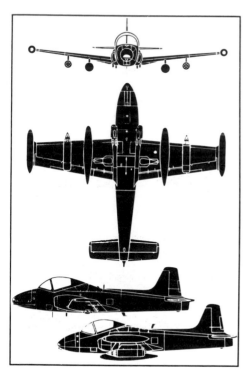

A two-seat basic trainer/light strike aircraft, the Strikemaster was developed from the Jet Provost, which was the RAF's first jet trainer. The Strikemaster has improved performance, more power and increased weapon load compared with the Jet Provost. The Jet Provost first flew in 1954 and the Strikemaster in 1967. The Strikemaster has a top speed of 760km/h and a weapon load of 1,360kg plus two fuselage-mounted machine guns. The Strikemaster has been widely exported. *Country of origin:* UK. *Main silhouette:* Strikemaster; *upper side view:* Jet Provost. *Picture:* Strikemaster.

Power: 1 × Viper turbojet   *Span:* 10.85m   *Length:* 10.64m

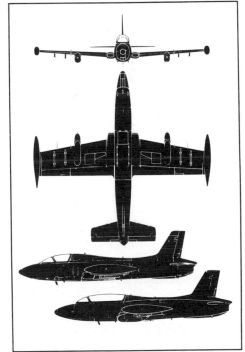

The MB-326 is a highly successful two-seat basic/advanced trainer and single-seat trainer/light attack aircraft (326K). Several hundred 326s of various versions have been built, and licence production has been undertaken in Australia and South Africa; in the latter country the type is called the Impala. Up to 1,815kg of armament can be carried on underwing pylons, including rockets, bombs, AS.12 missiles and gun pods. A reconnaissance pack can be fitted. Maximum speed is 686km/h. *Country of origin:* Italy. *Main silhouette:* MB-326; *upper side view:* MB-326K. *Picture:* MB-326K.

Aermacchi MB-339A
Confusion: Hawk, MB-326

Power: 1 × Viper turbojet   *Span:* 10.86m   *Length:* 10.97m

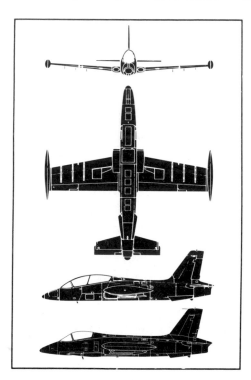

Based on and bearing a close resemblance to its predecessor, the MB-326, the two-seat MB-339A basic/advanced trainer has a redesigned forward fuselage. Auxiliary fuel tanks are carried at the wingtips and there are six underwing hardpoints for up to 1,815kg of bombs and rocket projectiles. Cameras or an armament pod can be carried under the forward fuselage. The prototype was flown in 1976 and the type has been produced for several air forces and navies. A single-seat tactical support variant is known as the MB-339K Veltro 2. Maximum speed is 898km/h. *Country of origin:* Italy. *Main silhouette:* MB-339A; *lower side view:* MB-339K Veltro 2. *Picture:* MB-339A.

Power: 1 × TFE731 turbofan   *Span:* 10.6m   *Length:* 12.5m

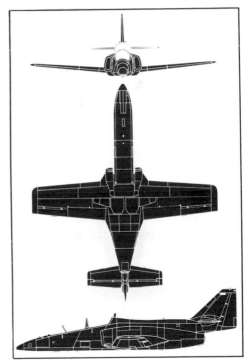

A two-seat basic/advanced jet trainer and light ground-attack aircraft, the C-101EB Aviojet went into service with the Spanish Air Force in 1980. In its armed export version the C-101BB can carry two machine guns or a cannon under the fuselage and there are six wing positions for external stores such as bombs, rockets, guided missiles or ECM pods. In light attack form the aircraft is designated C-101CC. Maximum speed is 775km/h and range 3,000km. *Country of origin:* Spain.

AIDC AT-3 *Confusion:* Aviojet

Power: 2 × TFE 731 turbofans   *Span:* 10.46m   *Length:* overall: 12.9m

A two-seat basic and advanced trainer attack aircraft built by Aero Industry Development Corp. in Taiwan, the AT-3 first flew in September 1980. Over 60 production AT-3s have been ordered for the Chinese Nationalist Air Force. The AT-3 has a maximum speed of 898km/h and an endurance of 3hr 12min. A weapons bay is situated under the rear cockpit, which can accommodate various stores including machine gun packs. There is one external stores station under the fuselage and two under each wing. *Country of origin:* Taiwan.

Power: 1 × TFE 731 turbofan   *Span:* 9.69m   *Length:* 10.93m

Sixty-four IA 63 two-seat tandem basic/advanced jet trainers have been ordered for the Argentine Air Force to replace the MS 760 Paris III. First flown in October 1984, deliveries of the IA 63 began in 1988. The West German firm of Dornier provided major technical assistance in the development phase and the IA 63 bears a close resemblance to a straight-wing Dassault-Breguet/Dornier Alpha Jet. Top speed of the IA 63 at sea level is 740km/h, maximum rate of climb 1,620m/min and service ceiling 12,900m. *Country of origin:* Argentina.

PZL I-22 Iryd

Confusion: Pampa

Power: 1 × SO-3W22 turbojet   *Span:* 9.6m   *Length:* 13.22m

A twin-engined shoulder wing trainer/light attack aircraft, the two seat I-22 Iryd (Irydium) first flew in March 1985. Capable of undertaking a variety of roles the I-22 carries one 23mm twin barrel cannon in an underfuselage pack plus underwing pylons for bombs, rockets or fuel tanks. Maximum speed at sea level is 915km/h and range, on internal fuel, 1,670km. *Country of origin:* Poland.

Power: 1 × IA-25TL turbofan   *Span:* 9.11m   *Length:* 12.11m

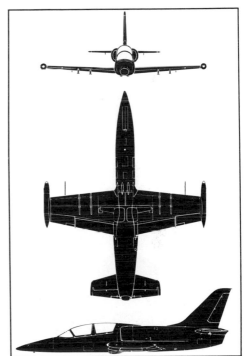

Successor to the L-29 Delfin, the tandem two-seat Albatros advanced trainer has engine intakes repositioned in the upper fuselage sides behind the cockpit. A single-seat light ground-attack/reconnaissance version, the L-39ZA, has four underwing pylons and a 23mm cannon in a pod mounted on the belly below the cockpit. The L-39V is the basic Albatros equipped with a target-towing winch for anti-aircraft artillery practice. The Albatros has a top speed of 610km/h. Deliveries began in 1973 and production (including exports) has already totalled well over 2,000 units. The L-39 was selected as the standard jet trainer of all the Warsaw Pact countries with the exception of Poland. An improved and up-rated version is designated L-39MS. *Country of origin:* Czechoslovakia. *Silhouette and picture:* L-39.

Northrop T-38A Talon *Confusion:* F-5, Starfighter, Hornet

Power: 2 × J85 reheated turbojets   *Span:* 7.7m   *Length:* 14.13m

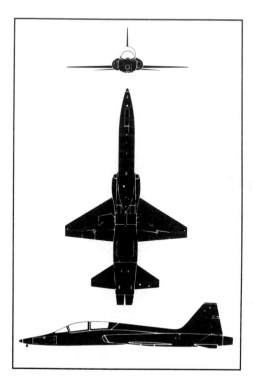

Very similar in outline to the Northrop F-5, the two-seat T-38A Talon was in fact a separate development with a different structure. The Talon was the first supersonic advanced trainer to be put into production, entering service with the USAF in 1961. A total of 1,187 Talons were built before production finished in 1972. The Talon has a maximum range of 1,759km. *Country of origin:* USA.

Power: 2 × J85 reheated turbojets   *Span:* 8.13m   *Length:* 14.68m

The most successful lightweight jet combat aircraft ever built, the Northrop F-5 flew for the first time in 1959. It is used for interception, ground attack and reconnaissance. The F-5A is a single-seater with four wing pylons and two 20mm nose guns, the RF-5A is the reconnaissance variant and F-5B is a two-seater trainer. The F-5E Tiger II air-superiority fighter has increased power, improved nav-attack system and a modified wing with leading-edge root extensions. Tiger II has a maximum speed of 1,732km/h and a range of 3,175km. The reconnaissance version of Tiger II is designated RF-5E Tiger Eye (with extended nose) and the two-seat version F-5F. *Country of origin:* USA. *Main silhouette:* F-5A; *first upper view:* F-5B; *second upper view:* F-5E; *top side view:* RF-5E. *Picture:* CF-5A.

Lockheed F-104 Starfighter   *Confusion:* F-5, Hornet

Power: 1 × J79 reheated turbojet   *Span:* 6.68m   *Length:* 16.69m

A single-seat all-weather tactical strike and reconnaissance fighter, the Lockheed F-104 Starfighter first flew in 1954 and was built in very large numbers. Unlike its swept-wing contemporaries, the Starfighter was designed with a very thin, straight wing. Air-to-air missiles or auxiliary fuel tanks can be carried on the wingtips, while a variety of missiles, bombs or fuel tanks can be accommodated on four underwing pylons. Standard fuselage-mounted armament is a 20mm rotary cannon. Top speed at altitude is 2,124km/h or Mach 2. Main service variant is the multi-role F-104G. The F-104S, licence-built in Italy, carries two Sparrow medium-range air-to-air missiles. The tandem two-seat trainer version, shown in the upper side view, is designated TF-104G. *Country of origin:* USA. *Main silhouette:* F-104S. *Picture:* F-104G.

McDonnell Douglas/Northrop F/A-18A Hornet

Power: 2 × F404 reheated turbofans   *Span:* 11.43m   *Length:* 17.07m

Based on a Northrop private-venture design, the F-18 Hornet is a single-seat carrierborne air-superiority fighter with Mach 1.8 performance and high manoeuvrability. McDonnell Douglas is the prime contractor for the Navy and Marine Corps versions F/A-18A fighter, A-18 attack aircraft and TF/A-18A two-seat trainer, while Northrop is a co-producer. In fighter form the Hornet carries a six-barrel 20mm cannon plus air-to-air missiles and other stores on nine wing positions, including the wingtips. The Canadian version is known as CF-18 and that for Spain, EF-18. The first Hornet unit became operational in 1983. *Country of origin:* USA. *Silhouette:* F/A-18A. *Picture:* CF-18.

Sukhoi Su-25 'Frogfoot'

Confusion: Citation, Magister

Power: 2 × Tumansky turbojets   *Span:* 14.3m   *Length:* 15.4m

The Soviet counterpart to the American A-10 Thunderbolt, the Su-25 'Frogfoot' is a large shoulder-wing, single-seat, twin-engined ground attack aircraft. It was used in action in Afghanistan and in Iraq. The type is also in service with the Czech Air Force and Iraq. Top speed is about 980km/h, radius of action 556km and up to 400kg of bombs can be carried on eight underwing pylons. A single cannon is carried under the fuselage. A two-seat trainer version is the Su-25 UB. *Country of origin:* USSR.

Aérospatiale Magister

Power: 2 × Marboré turbojets   *Span:* 11.4m   *Length:* 10.06m

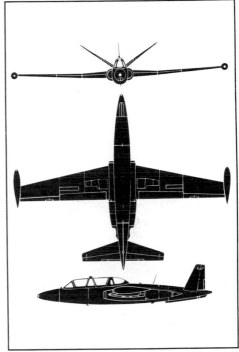

Nearly 1,000 CM170 Magister two-seat basic trainers were built before production ceased. First flown in 1951, the Magister has a distinctive layout, with high-aspect-ratio wings and V-tail. The CM170-2 is known as the Super Magister. Top speed is 700km/h and endurance 2hr 40min. Two machine guns are mounted in the nose, and rockets, bombs or missiles can be carried under the wings. The Magister has been used as a light ground attack aircraft, serving with the Israeli Air Force in the Middle East. A naval version known as the Zephyr was produced for shipboard carrier familiarisation duties by the French Navy. *Country of origin:* France. *Silhouette and picture:* Magister.

Rockwell International T-2 Buckeye

Confusion: Iskra

Power: 2 × J60 or J85 turbojets   *Span:* 11.62m   *Length:* 11.67m

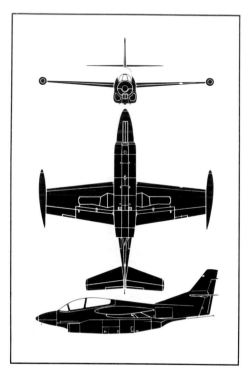

A shipborne basic trainer/attack aircraft, the T-2 Buckeye first flew, as the T2J-1 (T-2A) with one J34 engine, in 1958. The T-2B has two J60 engines and the T-2C and D two J85s. The wing was based on that of the early FJ-1 Fury fighter. Armament is carried on underwing stations. Maximum speed is 840km/h and range 1,685km. Buckeyes were exported to two countries: Venezuela (24 examples) and Greece (40). *Country of origin:* USA. *Silhouette and picture:* T-2C.

Power: 2 × VK-1 turbojets   *Span:* 21.45m   *Length:* 17.65m

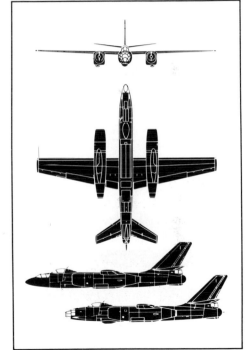

Although it is nearly 40 years since the Il-28 bomber, code-named 'Beagle', entered service, it is still in use with various air forces in Africa, Asia and South-east Asia and in small numbers in Eastern Europe. The trainer version, the Il-28U, has a second, lower cockpit and is code-named 'Mascot'. In bomber form the 'Beagle' carries 1,000kg of bombs internally and two 23mm cannon in the tail. Maximum speed is 900km/h and maximum range 2,180km. A Chinese-built version is known as the Harbin H-5. *Country of origin:* USSR. *Main silhouette:* Il-28; *upper side view:* Il-28U. *Picture:* H-5.

BAe Canberra   *Confusion:* 'Beagle', B-57

Power: 2 × Avon turbojets   *Span:* 19.5m   *Length:* 19.96m

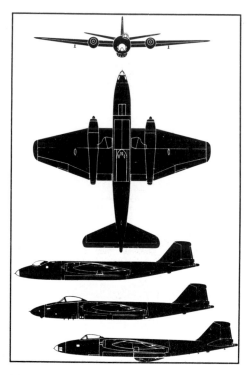

Still in widespread use 40 years after its first flight, the Canberra is used for tactical bombing, reconnaissance, training, electronic warfare and target towing. Production totalled 1,461, including examples licence-built in the USA (as the B-57) and Australia. Later versions have a raised offset cockpit. Maximum speed is 930km/h and range 6,100km. *Country of origin:* UK. *Main silhouette:* B(I)8; *top side view:* B6; *middle side view:* PR9. *Picture:* TT18.

Power: 1 × J57 or 1 × J75 turbojet   *Span:* 24.38m   *Length:* 15.11m

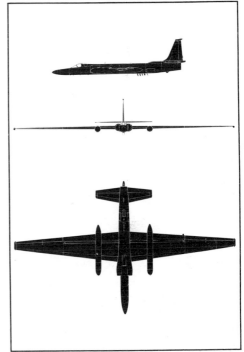

A high-altitude special-purpose reconnaissance aircraft, the single-engined U-2 came to the world's attention in 1960 when one was shot down over the Soviet Union. A very-high-aspect-ratio wing gave the U-2 a service ceiling of about 24,384m; cruising speed is 740km/h and endurance over eight hours. The production line was reopened to turn out 12 much modified U-2s (U-2R) and 35 aircraft with improved sensors and designated TR-1A and B, the B being a trainer with an additional cockpit. One example was supplied to NASA as the ER-2. *Country of origin:* USA. *Silhouette and picture:* TR-1A.

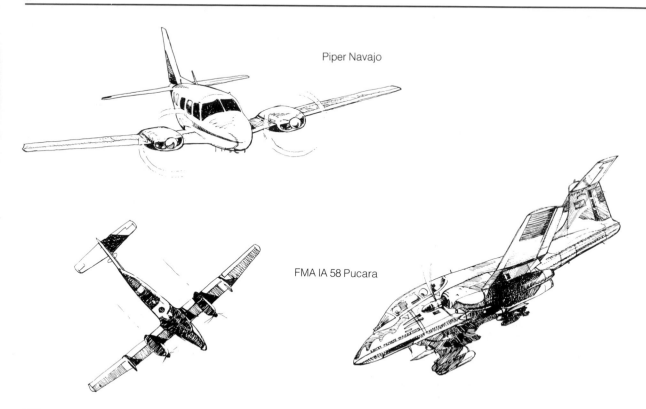

Piper Navajo

FMA IA 58 Pucara

Power: 2 × Gipsy Queen piston engines   *Span:* 17.40m   *Length:* 11.96m

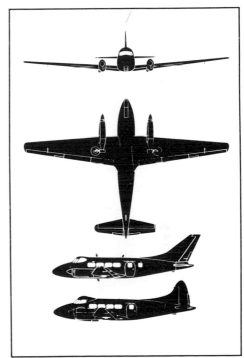

A total of 534 Dove 8/11-seat light transports were built over nearly 20 years. Series numbers were allotted largely to signify power increases in the Gipsy Queen engines. The military version of the Dove, the Devon, remains in service in the UK and Sri Lanka. The Royal Navy uses the name Sea Devon. Economical cruising speed is 310km/h and range 1,415km. A small number of Doves were re-engined by Riley and fitted with a swept fin and rudder. *Country of origin:* UK. *Main silhouette:* Dove; *upper side view:* Riley Dove. *Picture:* Dove.

Fairchild Metro

Confusion: ST-27

Power: 2 × TPE331 turboprops   *Span:* 17.37m   *Length:* 18.09m

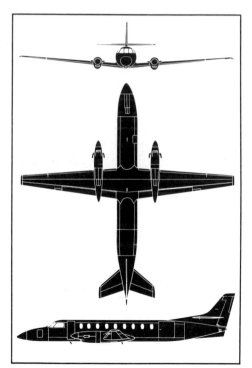

Developed from the Merlin, the Metro is a 20-passenger commuter airliner with a range of 804km. The Metro II has deeper windows than the Metro, and both versions are characterised by long noses. An executive transport version of the Metro II is designated Merlin IVA and has seating for 12/15 passengers. The Metro III and IIIA have a new, longer-span wing, while the the IIIA also has PT6A engines. The III has a range of 1,149km. Introduced in 1981, the Metro IIIC incorporates the wing introduced on the Metro III, as well as the latter's streamlined nacelles and new main landing gear doors. The cargo version of the Metro III is called Expediter while the Merlin IVC is a corporate version. C-26A is a variant with the US Air National Guard. *Country of origin:* USA. *Silhouette:* Metro IIIA. *Picture:* Merlin C-26A.

Power: 2 × Bastan turboprops   *Span:* 19.59m   *Length:* 15.3m

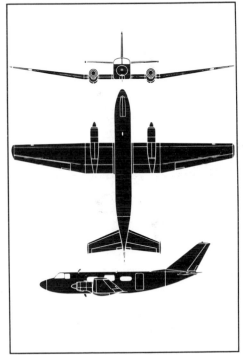

Developed from the Huanquero, the IA 50 Guarani seats up to 15 passengers or can be used for navigation training or air ambulance work. Compared with the Huanquero, the Guarani has a swept single fin. Both types can carry tip tanks. Cruising speed is 450km/h and range 1,995km. First flown as the Guarani I, embodying 20 per cent of the structural components of the Huanquero, this type was developed into the Guarani II, the definitive standard. The Guarani II features more powerful engines, de-icing equipment, a single swept fin and rudder in place of the Guarani I's twin-fin arrangement, and a shorter rear fuselage to save weight. The type was phased out of production in 1973. *Country of origin:* Argentina. *Silhouette and picture:* Guarani II.

Convair CV-240/340/440 Metropolitan

Confusion: DC-3, Guarani, 4-0-4, Crate

Power: 2 × R-2800 piston engines   *Span:* 32.12m   *Length:* 24.14m

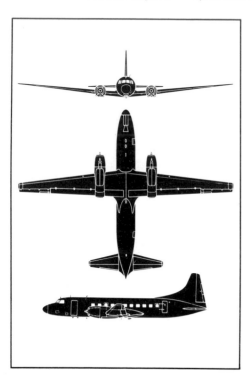

Originally built as the 40-seat Convair 240, the Metropolitan was developed into a number of variants. The 340 had a lengthened fuselage seating 44 passengers, while the 440 ultimately had accommodation for up to 52. Versions were also produced for the US Air Force and US Navy for air ambulance work, transport, crew training and ECM training. Altogether 1,081 240/340/440s were completed. The 440 cruises at 465km/h for 2,092km. *Country of origin:* USA. *Silhouette:* CV-340. *Picture:* CV-440.

252

Power: 2 × Allison 501 or 2 × Dart turboprops   *Span:* 32.12m   *Length:* 24.14m

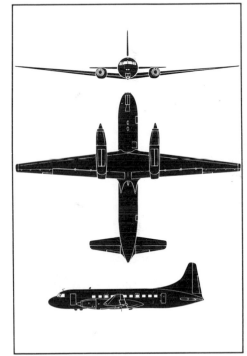

The Convair 340 and 440 both proved capable of conversion to turboprop power. Pacific Automotive converted more than 115 to Allison 501 turboprops (CV-580), while kits were produced for refitting 340/440s with Rolls-Royce Darts to produce the CV-600/640. The 640 with Darts can carry 56 passengers for 1,975km at a speed of 482km/h. *Country of origin:* USA. *Silhouette and picture:* CV-640.

Power: 2 × R2800 piston engines   *Span:* 28.44m   *Length:* 22.75m

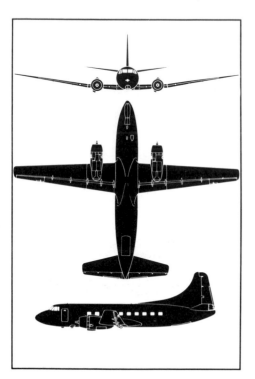

Developed from the Martin 2-0-2, which first flew in 1946, the 4-0-4 pressurised twin-engined transport entered service in 1951. The type was a rival to the Convair 340/440, which it resembles in outline. Over one hundred 4-0-4s were built and a number remain in use in the USA and South America. Forty passengers are carried, cruising speed is 448km/h and range is 1,730km. Other performance figures include a maximum speed of 500km/h and service ceiling of 8,845m. *Country of origin:* USA.

Power: 2 × Ash-82 piston engines   *Span:* 31.69m   *Length:* 22.3m

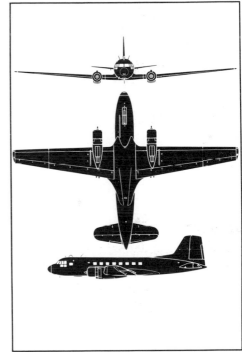

Some 3,600 Il-14 'Crate' transports were built for both civil and military use. The civil version seats 18/26 passengers, while the military 'Crate' has port-side double freight doors. A stretched version, the Il-14M, carries 24/28 passengers. A line in Czechoslovakia produced a 32-passenger variant, the Avia 14-32, the freighter 14T and the photographic survey 14FG. In 1960 the Czechs produced a pressurised 42-passenger 'Crate' with circular windows known as 14-42. The 14M has a cruising speed of 310km/h and a range of 1,304km. The 'Crate' is still in service in the Eastern Bloc and was widely exported. *Country of origin:* USSR. *Silhouette and picture:* Il-14M.

Douglas DC-3/Dakota

Confusion: 'Crate', Commando, Azor

Power: 2 × R-1830 piston engines   *Span:* 28.96m   *Length:* 19.63m

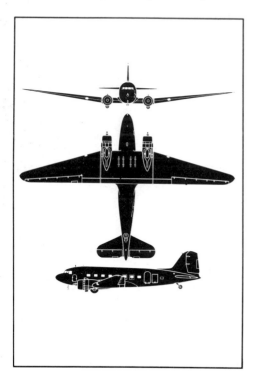

Probably the most famous of all airliners/military transports, the DC-3/Dakota first flew in 1935 and nearly 13,000 were built, including some 2,000 in Russia and Japan. Several hundred DC-3/Dakotas are still in use and there have been many variants, including turboprop developments. The basic DC-3 seats up to 36 passengers and cruises at 312km/h for 2,430km. The C-47 Dakota military version is in use in over 40 countries. *Country of origin:* USA. *Silhouette:* DC-3. *Picture:* C-47.

Curtiss C-46 Commando

Power: 2 × R-2800 piston engines   *Span:* 32.92m   *Length:* 23.26m

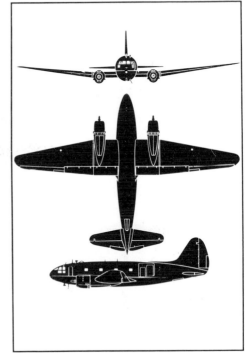

Like the ubiquitous DC-3/Dakota, the Curtiss Commando twin radial-engined transport has remained in service continuously since the Second World War. The type is in civil and military use, carrying either passengers or freight. Passenger versions can carry up to 62 passengers for 1,880km at a cruising speed of 301km/h. A total of 3,180 Commandos were built, of which a number are still operating, particularly in the Caribbean and South America. *Country of origin:* USA.

257

Dassault-Breguet Atlantic

Confusion: Commando, 'Crate', Neptune

Power: 2 × Tyne turboprops   *Span:* 36.3m   *Length:* 31.75m

A multi-national European product, the Atlantic long-range maritime patrol aircraft is in service in France, West Germany, Italy, the Netherlands and Pakistan. First flown in 1961, the Atlantic has a 12-man crew and can cruise at 320km/h for 18hr. The weapon load of acoustic torpedoes, depth charges or mines is carried internally. An updated version is the Atlantique 2, (ATL2), previously known as the Atlantic NG (*Nouvelle Génération*), with improved systems and structure, and capable of carrying Exocet missiles in the weapons bay. The French Navy has a requirement for 42 Atlantique 2s. *Country of origin:* France. *Silhouette:* Atlantic Mk 1. *Picture:* Atlantic 2.

Kawasaki/Lockheed P-2 Neptune

Power: 2 × T64 turboprops plus 2 × J3 turbojets　　*Span:* 29.78m　　*Length:* 29.23m

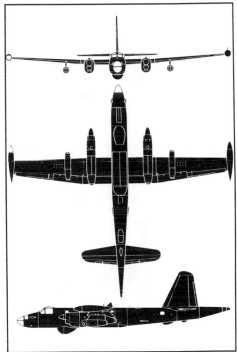

Originally designed and built by Lockheed in America, the P-2 Neptune maritime patrol and anti-submarine aircraft was subsequently manufactured under licence in Japan. After completing a number of P-2H Neptunes, Kawasaki developed a longer, higher-powered and longer-range version, the P-2J. Carrying 12 crew, the P-2J cruises at 370km/h and has a range of 4,450km. Up to 3,630kg of weapons can be carried in the internal bomb bay, while rocket projectiles may be fitted under the wings. The P-2J's operational equipment is known to be comparable in standard to that carried by the P-3 Orion and to include APS-80J search radar, exhaust-gas detector, and magnetic anomaly detector (MAD) in the tail boom. A total of four P-2Js have been converted to UP-2J target tugs. *Countries of origin:* Japan/USA. *Silhouette and picture:* P-2J.

Grumman OV-1/RV-1 Mohawk   *Confusion:* —

Power: 2 × T53 turboprops   *Span:* 14.63m   *Length:* 12.50m

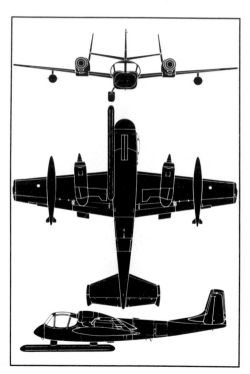

Developed as a US Army battlefield surveillance aircraft, the OV-1 Mohawk carries a wide range of cameras and electronic sensors, and can operate from small, rough sites. OV-1A has cameras, OV-1B sideways-looking radar (SLAR), OV-1C cameras and infra-red sensors, and OV-1D cameras and sideways-looking radar or infra-red. The RV-1D is a conversion for electronic intelligence work. The SLAR is mounted in a long, rectangular under-fuselage pod which extends well forward of the nose. Large auxiliary fuel tanks are carried under the wings. The crew of two are seated side by side and maximum speed is 491km/h. *Country of origin:* USA. *Silhouette:* OV-1B. *Picture:* OV-1D.

Power: 2 × M332 piston engines   *Span:* 12.25m   *Length:* 7.77m

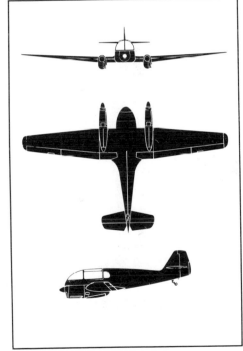

The original Aero 4/5-seat light twin was flown in 1947. It was followed by the Super Aero 45 and later by the 145 with uprated engines. The 45/145 remained in production until 1961, by which time some 700 had been built, of which over 600 had been exported. A particular recognition point is the streamlined, unstepped nose. The 145 cruises at 250km/h for up to 1,700km. Empty weight is 960kg and normal loaded 750kg. *Country of origin:* Czechoslovakia. *Silhouette and picture:* Aero 145.

Beechcraft Baron

Confusion: Aero 145, Navajo, Apache

Power: 2 × Continental piston engines   *Span:* 11.53m   *Length:* 9.09m

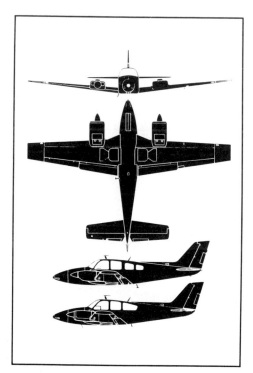

The B55 Baron four-passenger light transport was developed from the D95 Travel Air, which was itself a scaled-down Twin Bonanza. First flown in 1960, the Baron has been steadily updated over the years and has appeared under a variety of designations: A55, B55, C55, D55, E55, Turbo Baron and Baron 58/58TC. A pressurised variant is called the Baron 58P; over 2,200 Barons of all types have been delivered. Maximum cruising speed is 370km/h and range 1,950km. *Country of origin:* USA. *Main silhouette:* D55; *upper side view:* Baron 58. *Picture:* Baron 58.

Piper PA-31 Navajo

Power: 2 × Lycoming piston engines   *Span:* 12.4m   *Length:* (-31C): 9.94m

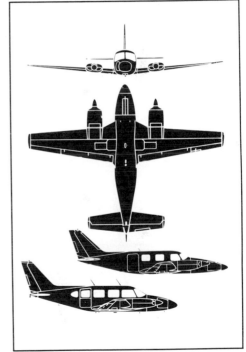

There have been several variants of the PA-31 Navajo since it was first introduced in 1964. Seating 6/8, the Navajo is used as an executive aircraft and a commuter airliner. When fitted with turbo-supercharged engines the type is called the Turbo-Navajo. The PA-31P, produced in 1970, has a pressurised cabin with fewer windows, and a longer fuselage. The PA-31C can fly 1,973km at a cruising speed of 383km/h. *Country of origin:* USA. *Main silhouette:* PA-31; *upper side view:* PA-31P. *Picture:* PA-31.

Piper PA-23 Apache

Confusion: Baron, Navajo

Power: 2 × Lycoming piston engines   *Span:* 11.33m   *Length:* 8.41m

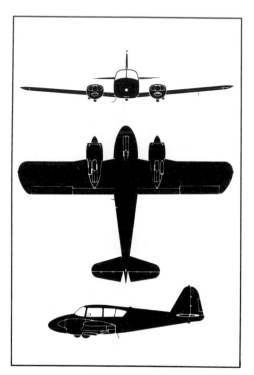

Originally known as the Twin-Stinson, the Piper PA-23-160 Apache first flew in 1952. In this form it had a curved fin, rudder and tailplane. Ten years later it was followed on the line by the PA-23-235 with a more bulbous cabin, a larger, square-cut fin and rudder, and square-cut tailplane. The Apache is similar to the Aztec but has 4/5 seats, a shorter nose and lower-powered engines. The 235 cruises at 307km/h for up to 1,900km. *Country of origin:* USA. *Silhouette:* PA-23-160. *Picture:* Apache 235.

Power: 2 × Lycoming piston engines   *Span.* 11.34m   *Length:* 9.52m

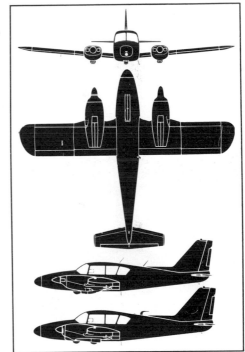

Originally built as the five-seat Aztec A in 1959, the Piper PA-23 evolved into the six-seat, longer-nosed Aztec B in 1962. The Aztec C, introduced in 1970, had redesigned engine nacelles and a longer nose, and Aztec D was externally similar. The latest version is the Aztec F. Aztecs of various marks are fitted with turbo-superchargers. As is the case with many American light aircraft, new designations often denote updated equipment and fittings rather than any change in outline. The Aztec F cruises at 338km/h for 1,335km. *Country of origin:* USA. *Main silhouette:* Aztec D; *lower side view:* Aztec B. *Picture:* Turbo Aztec F.

Gulfstream Aerospace GA-7 Cougar

Confusion: Cessna 310

Power: 2 × Lycoming piston engines   *Span:* 11.23m   *Length:* 9.1m

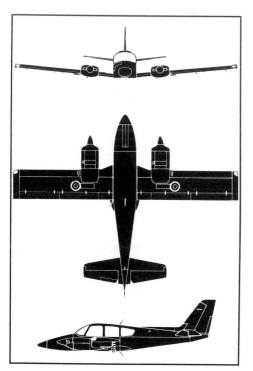

Originally a Grumman product, the GA-7 Cougar is intended for business and training use and can seat 4/6 people. First flown in 1974, the Cougar was significantly re-engineered before the production prototype flew in 1977. Gulfstream Aerospace's first entry into the lightweight twin-engined market, the Cougar was intended mainly for business use and for private pilots who already have instrument-flying experience. It can also be used as an economical twin-engined trainer. The production prototype flew for the first time on January 14, 1977, and delivery of production aircraft began in February 1978. Production has ceased. The Cougar has a maximum range of 2,035km and cruises at 305km/h. *Country of origin:* USA.

Power: 2 × Continental piston engines   *Span:* 11.25m   *Length:* 9.74m

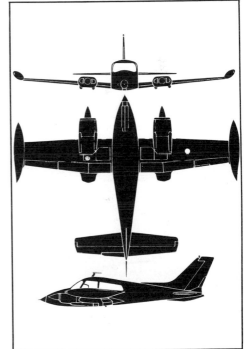

First flown in 1953, the Cessna 310 four-seat twin had a straight fin in its original form. In addition to hundreds of civil sales, the type was adopted by the USAF as the U-3A. With engine improvements incorporated, the 310 was marketed as the Riley 65 Rocket and Turbo Rocket. In 1960 the Cessna Model 310 appeared with a swept fin and more windows. Over 5,000 310s were built and a large number of variants are flying, including some with turbo-supercharged engines and a further USAF version, the U-3B. Another development is the 320C Skyknight, seating up to seven. Cruising speed is 267km/h and range is up to 1,141km. *Country of origin:* USA. *Silhouette and picture:* Cessna 310.

Piper PA-30/PA-39 Twin Comanche

Confusion: Seneca

Power: 2 × Lycoming piston engines   *Span:* 11.22m   *Length:* 7.67m

Developed as a twin version of the PA-24 Comanche single-engined light aircraft, the Twin Comanche first flew in 1961 and has been built in large numbers. A four/six-seater, it cruises at 319km/h and has a range of 1,335km. There are many variants, differing mainly in engine power, propellers and internal layout. Some are fitted with wingtip fuel tanks. From 1971 the PA-30 was succeeded by the externally similar PA-39. Refinements introduced with the later Twin Comanche variants include turbo-supercharging, first applied to the Turbo Twin Comanche B and C, and counter-rotating propellers (to eliminate torque effects) on the Twin Comanche C/R. *Country of origin:* USA. *Silhouette:* PA-39. *Picture:* PA-30C.

Power: 2 × Continental piston engines   *Span:* 11.85m   *Length:* 8.73m

The PA-34 Seneca six-passenger light transport and trainer was introduced in 1971. It has twin turbo-supercharged Continental engines, cruises at 285km/h and has a range of 1,635km. Poland holds a licence to build the Seneca II and distribute it in Eastern Europe. These aircraft are powered by PZL-Franklin engines and designated PZL-112 M-20 Mewa (Gull). Latest version of the PA-34 has detail modifications and is called the Seneca III. *Country of origin:* USA. *Silhouette:* Seneca II. *Picture:* Seneca III.

Cessna 340A *Confusion:* Duke, Aerostar

Power: 2 × Continental piston engines   *Span:* 11.6m   *Length:* 10.46m

Developed from the Cessna 310, the Model 340A is a four-passenger pressurised business aircraft. The layout is similar to that of the 310 but the fuselage cross-section is different and the windows are circular. The 340 in effect replaced the 320C Skyknight. As with other similar types in the Cessna range, the powerplants are Continental flat-four piston engines, which have a distinctive rectangular appearance in the head-on view. Range of the 340A is 2,543km at an economical cruising speed of 332km/h. Later 340A variants feature a number of improvements, including better forward view as a result of structural changes, and a new strobe lighting system. *Country of origin:* USA.

Power: 2 × Lycoming piston engines *Span:* 11.18m *Length:* 10.61m

The first of the Aerostar series of light transport aircraft flew in 1967 as the Model 600. The 601P has more wingspan and a pressurised fuselage. A six-seater, the Aerostar 601 is one of a small number of types with a straight wing and swept tail surfaces. The 601B cruises at 434km/h and has a range of 2,309km. Around 1,000 Aerostars of all types have been sold. The 700P has higher powered engines. Introduced in 1981, the Aerostar 602P is generally similar to the 601P. *Country of origin:* USA. *Silhouette:* 601P. *Picture:* 602P.

Cessna T303 Crusader

Confusion: Duke, Chancellor

Power: 2 × Continental piston engines   *Span:* 11.48m   *Length:* 9.27m

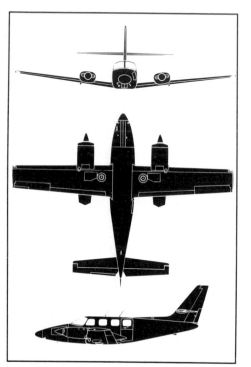

First flown in 1978 as the four-seat Model 303, the Crusader was redesignated T303. With six seats and turbo-supercharged engines, the Crusader was the first completely new twin-engined type to be put into production by Cessna for a decade. Deliveries began in September 1981 and 297 had been delivered by the time production was suspended at the end of 1987. Cruising speed is 333km/h and maximum range 1,861km. *Country of origin:* USA.

Power: 2 × Lycoming piston engines *Span:* 11.96m *Length:* 10.31m

A pressurised, turbo-supercharged light transport, the Beech B60 Duke seats 4/6 passengers, cruises at 431km/h and has a range of 1,872km. A total of 596 Dukes were built in two versions, the A60 and B60, before production ceased. The B60, powered by Lycoming TIO-541-E1C4 engines, entered production in 1974. *Country of origin:* USA. *Silhouette and picture:* B60.

Beechcraft Queen Air/Seminole

Confusion: King Air

Power: 2 × Lycoming piston engines   *Span:* 15.32m   *Length:* 10.82m

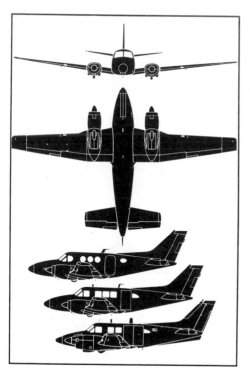

Like the rest of the Beech range, the Queen Air has been built in very large numbers for both civil and military use. A four/nine-passenger twin, the Queen Air has been produced as the 65/A65, U-8F (US Army), Model 70/80/A80/B80 and U-21A, B, C, E and G (US Army). The U-8F, characterised by a straight fin, is called the Seminole. The Queen Air is in service both as an executive aircraft and a feederliner. The B80 seats 11 passengers and is called the Queen Airliner. It has a maximum cruising speed of 360km/h and a range of 2,494km. *Country of origin:* USA. *Main silhouette:* B80; *top side view:* B88; *bottom side view:* U-21. *Picture:* B80.

Power: 2 × **PT6A turboprops**   *Span:* (100): **13.98m**   *Length:* **12.1m**

Based on the 65-80 Queen Air but with a redesigned pressurised fuselage, the King Air is a 16-seat twin-turboprop business aircraft. Models include the 90/A90/B90/C90/E90/100/B100 and T-44 advanced trainer for the USAF. Latest version is the C90A. Over 1,400 King Air 90s have been built and the type is in use in many countries. Maximum cruising speed is 460km/h and range 2,425km. *Country of origin:* USA. *Silhouette:* B100. *Picture:* E90.

Beechcraft B99 Airliner *Confusion:* King Air, Queen Air

Power: 2 × PT6A turboprops   *Span:* 14m   *Length:* 13.58m

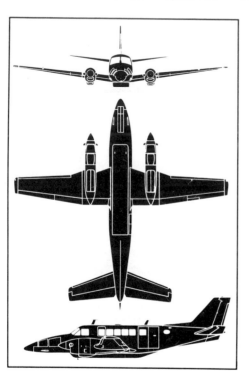

The B99 Airliner is a twin-turboprop unpressurised 17-seat feederliner/air taxi. It first flew in 1966 and deliveries began in 1968. Similar in outline to other Beech twins of similar size, the B99 has a particularly long nose. A main cargo door allows the aircraft to be used for all-cargo or passenger/cargo operations. Maximum cruising speed is 454km/h and range 853km. Offering increased power and systems refinements, the C99 was known as the C99 Airliner. Deliveries began in 1981. *Country of origin:* USA.

Power: 2 × TPE331 turboprops   *Span:* 15.04m   *Length:* 11.89m

Designed to fill a slot between piston-engined and turbofan business aircraft, the Cessna 441 Conquest II seats eight/ten passengers in a pressurised cabin. It has a high-aspect-ratio wing, while the general layout is similar to that of other Beech twins. Range is 2,077km and maximum cruising speed 513km. The prototype Conquest flew in 1975 and 360 had been produced when production was suspended in December 1987. *Country of origin:* USA.

Piper PA-31-350 Chieftain

Confusion: Conquest, Titan

Power: 2 × Lycoming piston engines   *Span:* 12.4m   *Length:* 10.55m

The PA-31-350 Chieftain has a longer fuselage than the Navajo and is a six/ten-seat executive/commuter/cargo aircraft. Cargo can be carried both in the cabin and in the nose. The Lycoming engines have turbo-superchargers. Cruising speed is 404km/h and range 1,640km. Production was suspended in June 1987 when nearly 2,000 Chieftains had been sold. *Country of origin:* USA.

Power: 2 × Continental piston engines   *Span.* 14.12m   *Length:* 12.04m

Intended to provide greater payload/range performance than the Cessna 402, the Cessna Titan was originally known as the 404. The cabin is convertible to cargo, feederliner and executive configurations and double doors can be fitted for loading large cargo. The all-passenger Titan is called the Ambassador, while the utility passenger/cargo version is known as the Courier. There are no significant external differences between the two variants. Maximum cruising speed is 369km/h and range 2,572km. Several hundred Titans were delivered. *Country of origin:* USA.

Cessna 421 Golden Eagle

Confusion: Cessna 401

Power: 2 × Continental piston engines   *Span:* 12.53m   *Length:* 11.09m

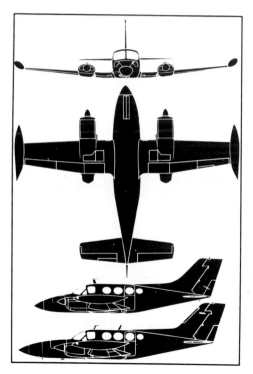

This series of pressurised business aircraft began in 1967 with the Model 421, which was developed into the 421B Golden Eagle and the 421B Executive Commuter. In 1975 came the 421C Golden Eagle, which had a new outer wing without the distinctive wingtip fuel tanks. The 421C Executive Commuter seats up to 11 as a feederliner and has a maximum cruising speed of 450km/h and a range of 2,317km. Later versions were Golden Eagle, Golden Eagle II and Golden Eagle III. A total of 1,909 Golden Eagles were delivered. *Country of origin:* USA. *Main silhouette:* Model 421A; *lower side view:* Model 421B. *Picture:* Model 421C.

Power: 2 × PT6A turboprops   *Span:* 13.45m   *Length:* 10.9m

A turboprop six/eight-seat pressurised business executive transport, the Conquest I (originally known as Corsair) is based on the airframe of the 421 Golden Eagle. Design began on November 1, 1977, and construction of a prototype was initiated three months later on January 30, 1978. This flew for the first time on September 12, 1978, and construction of a pre-production aircraft was started during 1979. By end 1987, 232 of the earlier Corsairs and the Conquest I had been delivered. At a cruising speed of 389km/h the Corsair has a range of 3,050km. A total of 499kg of baggage and up to six passengers can be carried. *Country of origin:* USA.

Cessna 401/402 *Confusion:* Cheyenne

Power: 2 × Continental piston engines   *Span:* 12.15m   *Length:* 11m

The Cessna 401 and 402 have basically similar airframes, the series having the names Utililiner and Businessliner for the six/eight-seat feeder and nine-seat executive versions. Several variants with differing cockpit, cabin and equipment standards have been built. Production of the 401 finished in 1972 and thereafter the 402 was standard, until production ceased at the end of 1987. The 402 Businessliner has a maximum range of 2,637km and a maximum cruising speed of 351km/h. Wingtip tanks are fitted as standard. *Country of origin:* USA. *Silhouette:* Model 401B. *Picture:* Model 402.

Piper PA-31T Cheyenne II/PA-31T2 Cheyenne II XL

Power: 2 × PT6A turboprops   *Span:* 13.01m   *Length:* 10.57m

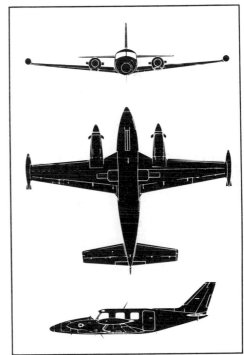

The first turboprop-powered Piper aircraft, the PA-31T Cheyenne II light transport first flew in 1969. The airframe is similar to that of the pressurised Navajo. Seating is provided for eight including the pilot. Large wingtip fuel tanks are fitted. Economical cruising speed is 393km/h and range is 2,739km. A lower-powered version is known as the Cheyenne I, while the Cheyenne II XL has a fuselage stretch. *Country of origin:* USA. *Silhouette:* Cheyenne II. *Picture:* Cheyenne II XL.

Cessna 414A Chancellor

Confusion: Conquest, Titan

Power: 2 × Teledyne Continental piston engines   *Span:* 13.4m   *Length:* 11.09m

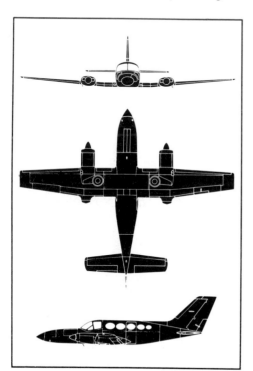

Cessna introduced the pressurised twin-engined Model 414 on December 10 1969, as a 'step-up' aircraft for owners of Cessna or other light unpressurised twins. It combined the basic fuselage and tail unit of the Model 421 with the wing of the Model 402 and had 310hp turbocharged engines. Successor to the 414, the 414A Chancellor is a utility transport seating up to six passengers in a pressurised cabin. Major changes from the Model 414 included a new 'wet' wing of increased span, and extended nose and baggage area. First introduced in 1978, the Chancellor has been built in I, II and III versions with differing equipment. Maximum speed is 443km/h and range at 267km/h is 2,340km. Over 1,000 414/414A aircraft were built before the line was suspended. *Country of origin:* USA.

Power: 2 × TPE331 turboprops   *Span:* 14.1m   *Length:* 12.85m

Produced in large numbers, the Swearingen Merlin series of executive transports has been steadily developed since its introduction in 1966. The original version was an eight-seater with two PT6 engines (IIA). Re-engined with TPE331s, this became the Merlin IIB. Merlin IIIA is an 8/11 passenger stretched version with four large side windows. The IIIB is powered by uprated TPE331s. The Merlin IIIC, was introduced in 1981 and was superceded in 1984, by the 8-10 passenger Fairchild 300 – similar to the IIIC but with winglets. *Country of origin:* USA. *Silhouette:* Merlin IIIC. *Picture:* Merlin 300.

Piper T-1040

Confusion: Chieftain, Cheyenne

Power: 2 × PT6A turboprops   *Span:* 12.5m   *Length:* 11.18m

A commuter airliner, the nine-passenger T-1040 combines the wings, nose and tail of the Cheyenne with the fuselage of the Chieftain and is powered by two 500 shp turboprops. The prototype T-1040 (T = transportation) first flew in July 1981 but only a small number were built. With full payload the T-1040 has a range of 724km, flying at a cruising speed of 441km/h. *Country of origin:* USA.

Power: 2 × TPE331 turboprops   *Span:* 15.85m   *Length:* 14.37m

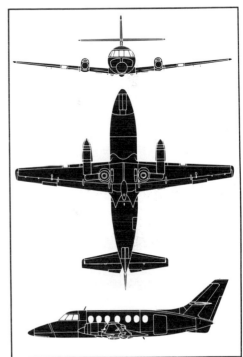

In its original form the Jetstream was developed by Handley Page and then Scottish Aviation. Thirty Astazou-powered Jetstreams 1 and 2 are still in civil service, while T1s and T2s are used by the RAF and the Royal Navy. British Aerospace Scottish Division (formerly Scottish Aviation) produced a much modified version called Jetstream 31, powered by two Garrett turboprops and aimed at the commuter and light business markets. A flight development aircraft flew in March 1980. Four Jetstreams, as T Mk 3 were ordered for the Royal Navy. By March 1989 some 252 Jetstreams had been sold. The Jetstream Super 31 has more powerful TPE 331 engines. A projected version is the Jetstream 41 which will carry 29 passengers as compared with 19 on the 31. *Country of origin:* UK.

Gulfstream Aerospace Gulfstream I/I-C

Confusion: Jetstream 31, BAe 748, YS-11A

Power: 2 × Dart turboprops   *Span:* 23.92m   *Length:* 19.43m

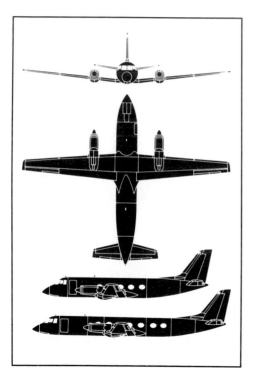

A long-range twin-engined business aircraft, the Grumman Gulfstream I seats up to 24 passengers. The layout is similar to that of the BAe 748, particularly in the engine/undercarriage nacelle arrangement. Maximum cruising speed is 560km/h and range with maximum fuel is 4,088km. Military versions include the TC-4C for the US Navy and the VC-4A for the US Coast Guard. Over 200 Gulfstream Is were built. The 37-passenger Gulfstream I-C commuter airliner, has a 3.25m fuselage extension. *Country of origin:* USA. *Main silhouette:* Gulfstream I; *lower side view:* Gulfstream I-C. *Picture:* Gulfstream I.

Power: 2 × **Dart turboprops** *Span:* **30.02m** *Length:* **20.42m**

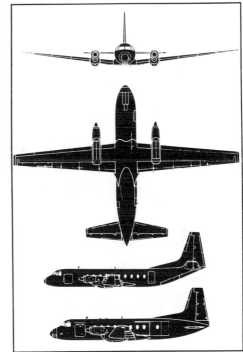

First flown in 1960, the 748 has ceased production and 380 examples were sold. Seating up to 58 passengers and with a maximum cruising speed of 448km/h, the 748 has been sold to many civil and military operators around the world. India has built a large number of 748s under licence. The Andover C1 cargo version has an upswept rear fuselage with loading doors, while some variants have a large aft side-loading door. Last civil version is the Super 748, with later mark Dart engines and internal refinements. *Country of origin:* UK. *Main silhouette:* 748 Military Transport; *upper side view:* Andover C1. *Picture:* Super 748.

 **NAMC YS-11A** *Confusion:* Gulfstream I, BAe 748, Saab-Fairchild 340, BAe ATP

Power: 2 × Dart turboprops   *Span:* 32m   *Length:* 26.3m

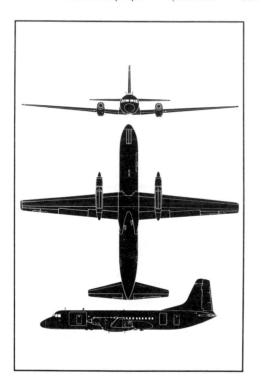

The YS-11A short-medium-range airliner went into service in 1965 and a total of 174 were built, including 22 for the Japan Maritime and Air Self-Defence Forces. The variants are designated -100, -200, -300, -400, -500 and -600. Up to 60 passengers can be carried at 466km/h for 1,090km. A big freight door can be fitted on the forward port side. The type has been exported and some are in service in the United States. *Country of origin:* Japan. *Silhouette:* YS-11A-300. *Picture:* JMSDF YS-11A.

Power: 2 × PW126 turboprops   *Span:* 30.63m   *Length:* 26m

To succeed the super 748, British Aerospace has developed the Advanced Turboprop (ATP). A 64 seat regional Airliner, the ATP first flew in August 1986 and by March 1989, 53 had been ordered. The ATP has the same cross section as the 748 but has a lower fuselage, wider span wings and swept tail. The Pratt and Whitney turboprops drive six blade propellers and the ATP is particularly quiet on take off and landing. The ATP cruises at 496km/h and has a maximum range of 3,444km. *Country of origin:* UK.

Saab 340A

Confusion: Gulfstream I, Bandeirante

Power: 2 × CT7 turboprops   *Span:* 21.44m   *Length:* 19.72m

Originally jointly developed with Fairchild as the Saab-Fairchild 340, this regional airliner is now a wholly Saab programme. A 34-seater intended for economical operation over short-haul, low-density routes, the 340 first flew in January 1983, and it entered service in June 1984. Over 100 orders and options for the type have been placed. The 340B is a higher powered version with increased tailplane span. The 340 cruises at 467km/h and has a range of 1,795km with maximum payload. *Country of origin:* Sweden.

EMBRAER EMB-110 Bandeirante

Power: 2 × **PT6A turboprops**   *Span:* 15.32m   *Length:* 15.08m

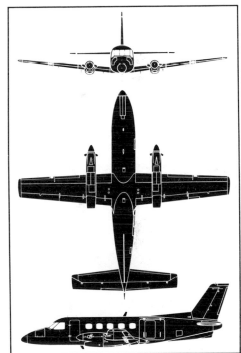

A very successful commuter transport, the EMB-110 Bandeirante seats 21 passengers (in 110P2 form), cruises at 360km/h and has a range of 1,916km. First deliveries went to the Brazilian Air Force in 1973 as the C-95. There are twelve versions of the Bandeirante, including the 110P1 (quick-change passenger transport), B1 (aerial photography), S1 (geophysical survey), K1 (military transport with longer fuselage), 110A (navigation/landing aid calibration), 110E (eight-seat executive transport) and 111 (radar-equipped maritime patrol aircraft). The EMB-110P1A/2A have dihedral tailplanes. *Country of origin:* Brazil. *Silhouette:* EMB-110K1. *Picture:* EMB-111.

Piper PA-44 Seminole

Confusion: Duchess

Power: 2 × Lycoming piston engines   *Span:* 11.75m   *Length:* 8.41m

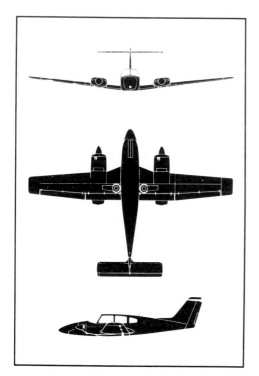

Bearing a close resemblance to the Beechcraft Duchess 76, the T-tail PA-44 Seminole is one of a new generation of American light aircraft intended for training and light transport duties. First flight was in 1976 and deliveries began in the spring of 1978. Although similar in appearance to other Piper aircraft, the Seminole was in fact a new design. Turbo Seminole is the name given to the turbo-supercharged version. Range is 1,546km and cruising speed 286km/h *Country of origin:* USA. *Silhouette:* Seminole. *Picture:* Turbo Seminole.

Power: 2 × Lycoming piston engines   *Span:* 11.59m   *Length:* 8.84m

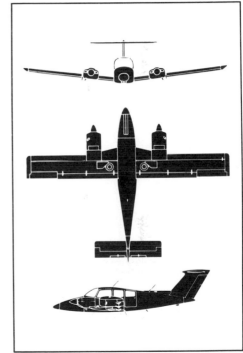

Embodying some parts from the single-engined Beechcraft Sierra, the Duchess 76 is a four-seat T-tail cabin monoplane designed for transport and training. The accent is on low cost and ease of production, and an unusual feature is the fitting of pilot doors on both sides of the aircraft. The Duchess first flew in production form in 1977. Several hundred have been delivered. Maximum cruising speed is 298km/h and range is over 1,290km. *Country of origin:* USA.

Piper PA-42 Cheyenne III/IIIA/400

Confusion: Super King Air 200, Duchess

Power: 2 × PT6A or 2 × TPE 331 turboprops   *Span:* 14.6m   *Length:* 11.58m

The PA-42 Cheyenne III first appeared in 1979 and differed from the rest of the Cheyenne family in having greater wingspan, longer fuselage, more windows, uprated engines and a prominent T-tail. A higher-powered version, the Cheyenne IIIA went into production in January 1984. Eight passengers are carried and cruising speed is 580km/h. A further variant, the Cheyenne 400 (formerly 400LS) flew in February 1983, powered by Garrett TPE 331 engines. *Country of origin:* USA. *Silhouette:* Cheyenne 400LS. *Picture:* Cheyenne IIIA.

Beechcraft Super King Air 200

Power: 2 × **PT6A turboprops**   *Span:* **16.61m**   *Length:* **13.34m**

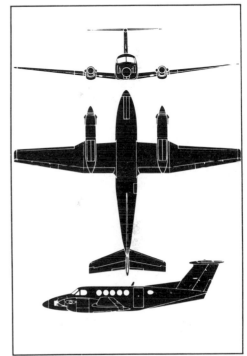

The Super King Air 200 represents a major departure in Beech twin design, with a wing of much higher aspect ratio and a T-tail. The prototype flew in 1972 and deliveries continue. In addition to civil applications as a feederliner/executive aircraft, the type has been ordered by the USAF, US Navy and US Army as the C-12A/C/D/E/F, UC-12B/F/M, RU-21J, RC-12D/H/K. A geographic survey version is the Model 200T, and a further variant is known as the Maritime Patrol B200T. The Super King Air 200 cruises at 503km/h and has a range of 3,495km. *Country of origin:* USA. *Silhouette and picture:* Super King Air 200.

Beechcraft 1900C

Confusion: King Air 200, Brasilia

Power: 2 × PT6A turboprops   *Span:* 16.61m   *Length:* 17.63m

First flown in September 1982, the Beechcraft 1900C twin-turboprop commuter airliner has been produced in three versions. The 1900C Airliner has a cargo door, the 1900C Exec-liner is a business aircraft and the C-12J is a miltary version serving with the US Air National Guard as a replacement for Convair C-131s. In addition, six aircraft have been procured by Egypt for electronic surveillance and maritime patrol. The basic 1900C carries two crew and up to 19 passengers. Maximum cruising speed is 495km/h and range, with 15 passengers, 2,383km. Particular recognition features are the tail-lets under the tail plane and the fixed horizontal tail surfaces on the rear fuselage forward of the tailplane. *Country of origin:* USA.

EMBRAER EMB-120 Brasilia

Power: 2 × PW115 turboprops   *Span:* 19.78m   *Length:* 20m

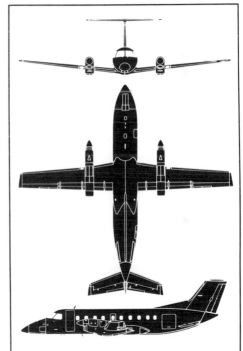

First flown in July 1983, the EMB-120 Brasilia is a twin turboprop passenger and cargo aircraft. Seating 30, the Brasilia has a cruising speed of 487km/h and a range of 1,112km. Deliveries commenced in mid-1985. Military versions include the C-97 and VC-97 for the Brazilian Air Force. By April 1988, 226 Brasilias had been ordered. *Country of origin:* Brazil.

EMBRAER EMB-121 Xingu

Confusion: Super King Air 200, Pucara, Brasilia

Power: 2 × PT6A turboprops   *Span:* 14.05m   *Length:* 12.25m

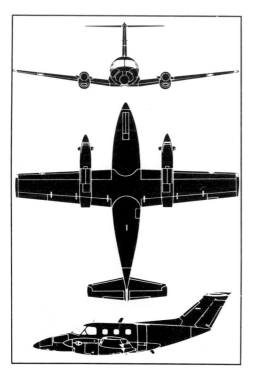

A pressurised five/six-passenger business aircraft, the EMB-121 Xingu first flew in 1976 and deliveries began in 1977. In addition to civilian sales, the Xingu has been bought by the Brazilian Air Force. Maximum cruising speed is 489km/h and range is 2,160km. The uprated Xingu II flew in 1981. A total of 105 Xingus were produced before the line closed in August 1987. *Country of origin:* Brazil. *Silhouette and picture:* Xingu I.

Power: 2 × **Astazou turboprops**   *Span:* 14.5m   *Length:* 14.25m

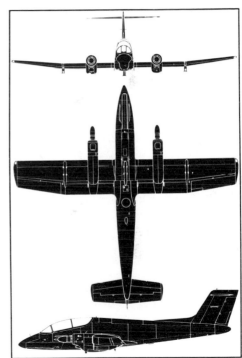

The IA 58 Pucara counter-insurgency (Coin) aircraft, first flew in 1969. Highly manoeuvrable, the Pucara is capable of using very short rough fields. The Pucara was widely used in the Falklands conflict in 1982. Two 20mm cannon and four machine guns are fitted in the fuselage and 1,500kg of bombs, rockets or missiles are carried on one underfuselage and two underwing pylons. The Pucara is in service with the Argentinian Air Force. Maximum speed is 500km/h and range is 3,042km. *Country of origin:* Argentina.

Beech 18 *Confusion:* Huanquero

Power: 2 × R-985 piston engines   *Span:* 15.14m   *Length:* 10.7m

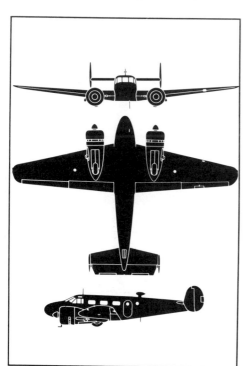

The original Beech 18 entered production in 1937 and several thousand had been completed by the end of the Second World War. Civil manufacture recommenced after the war and development led to the E18, G18 and H18 Super models, of which over 750 were built. Compared with early marks, the later 18s have extended rectangular wingtips, various nose lengths and changed window shapes. The H18 Super 18 carries nine passengers for 2,460km. Long-nose turboprop conversions of the Beech 18 by Hamilton Aviation are called the Westwind III and Westwind II STD. *Country of origin:* USA. *Silhouette:* Beech 18. *Picture:* H18 Super 18.

Let L-200 Morava

Power: 2 × M337 piston engines   *Span:* 12.31m   *Length:* 8.61m

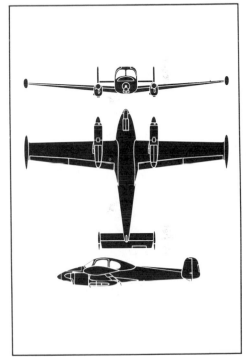

The successor to the Aero 145, the L-200 Morava is a four/five-seat light business aircraft first flown in 1957. Over 400 have been built and many have been exported within Eastern Europe. The three versions—the L-200, L-200A and L-200D—are all similar in external appearance. The Morava can be converted to carry two stretchers. Cruising speed is 256km/h and range 1,710km. *Country of origin:* Czechoslovakia. *Silhouette and picture:* L-200.

Twin propellers, high/shoulder wing

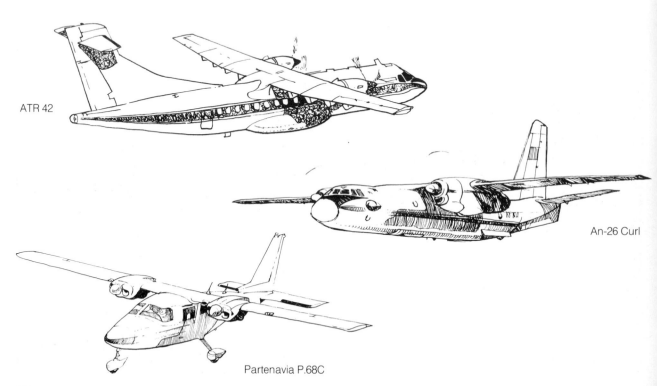

ATR 42

An-26 Curl

Partenavia P.68C

Power: 2 × Lycoming piston engines or 2 × TPE 331 turboprops   *Span:* 14.2m   *Length:* 13.1m

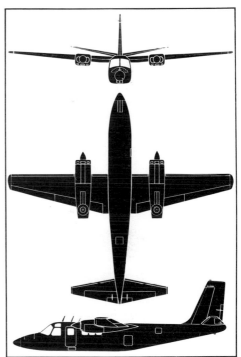

Beginning in 1948 with the Aero Commander, the Commander range is very extensive and many hundreds have been produced. From the 520 six-seater the design progressed to the Grand Commander with a longer fuselage and accommodation for nine passengers. Also in the series are the Shrike Commander four-seater, the shorter-wing, pressurised, eight-seat Hawk Commander, and a Garrett AiResearch turboprop version of the Hawk, the Commander Jetprop 840. Other versions are the Jetprop 900, 980 and 1000. The 840 has a range with maximum payload of 1,985km at 460km/h. *Country of origin:* USA. *Silhouette:* Aero Commander. *Picture:* Commander Jetprop 980.

Mitsubishi MU-2

Confusion: Turbolet

Power: 2 × TPE331 turboprops   *Span:* 11.94m   *Length:* 12.02m

A utility transport first flown in 1963, the MU-2 seats up to 11 passengers. Various versions, from the MU-2A to P, vary in power, weights and performance. The MU-2E is used by the Japan Air Self-Defence Force for rescue work. Wingtip tanks are standard. MU-2s were later assembled in the United States at a plant in Texas. The MU-2N has a cruising speed of 480km/h and a range of 2,330km. The MU-2N is called the Marquise and the MU-2P the Solitaire. Production ceased in March 1986 after 755 MU-2s had been built. *Country of origin:* Japan/USA. *Main silhouette:* MU-2B; *upper side view:* MU-2J. *Picture:* MU-2J.

Power: 2 × M601 or 2 × PT6A turboprops   *Span:* 17.48m   *Length:* 13.61m

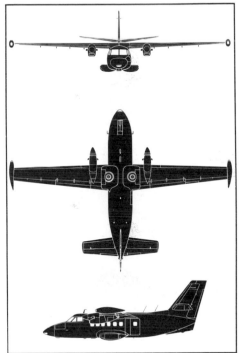

A light passenger/freight transport, the L-410 Turbolet first flew in 1969. Cruising at 380km/h for up to 200km, the Turbolet carries 17/19 passengers. The first production version, the L-410A, is powered by American PT6A engines, while the L-410M has two Czech M601A turboprops. A survey/photographic version, the L-410AF, has a large glazed nose on which is mounted a non-retractable nosewheel. Current production model is the L-410UVP-E, with increased span, larger fin and tailplane dihedral. The L-410UVP-E is fitted with wingtip tanks. *Country of origin:* Czechoslovakia. *Silhouette:* L-410UVP-E. *Picture:* L-410UVP.

Piaggio P.180 Avanti
Confusion: P.166

Power: 2 × PT6A turboprops   *Span:* 13.84m   *Length:* 14.17m

The P.180 Avanti is unusual in that it is virtually a triplane – the wing, tailplane and foreplane all generating lift. First flight of this five/nine passenger corporate transport took place in September 1986 and first deliveries were scheduled to take place in 1989. For three years Gates Learjet was a partner in the project but then withdrew. The Avanti has an economical cruising speed of 593km/h and range, with four passengers, of 3,335km. *Country of origin:* Italy.

Power: 2 × Lycoming piston engines or 2 × Lycoming turboprops   *Span:* 14.69m   *Length:* 11.9m

Produced in large numbers, the P.166 with two pusher engines was first flown in 1957. This light transport, seating up to nine passengers, has been built in both military and civil forms. Current versions are the P.166-DL2 with piston engines and P.166-DL3 with turboprops; the latter first flew in 1976. Wingtip tanks are fitted as standard. A maritime surveillance version is known as the P.166-DL3-MAR. Economical cruising speed of the -DL3 is 300km/h and range with maximum payload 741km. *Country of origin:* Italy. *Silhouette:* P.166-DL3. *Picture:* P.166M.

Beechcraft Starship 1 *Confusion: —*

Power: 2 × PT6A turboprops   *Span:* 16.6m   *Length:* 14.05m

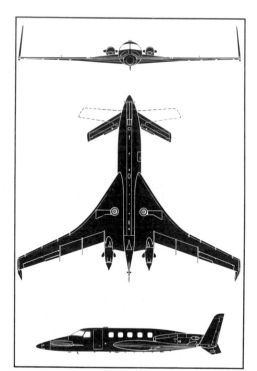

A very unusual canard design, the Beechcraft Model 2000 Starship 1 is an eight seat business aircraft made out of composite materials. Powered by two pusher PT6A turboprops, the Starship 1 has a crescent wing, endplate fins and variable geometry foreplanes. First flight of a full scale prototype was in February 1986 and first delivery was due in mid-1989. Over 50 have been ordered. Maximum take-off weight is 6,350kg, economical cruising speed 500km/h and range 4,032km. *Country of origin:* USA.

Power: 2 × TPE331 turboprops   *Span:* 16.97m   *Length:* (-200): 16.55m

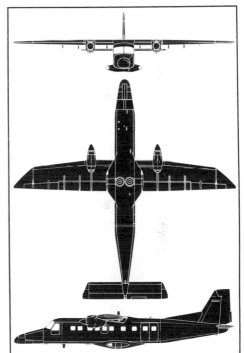

The Dornier Do 228 commuter transport is a combination of a new-technology wing with the modified fuselage of the Do 128 and a retractable undercarriage. The basic Do 228-100 has a length of 15.03m and seats 15 passengers, while the -200 is longer and seats 19. The 228-100 went into service in Norway in the summer of 1982, following first flights by the -100 and -200 in 1981. The 228 is being licence-built in India. There are two maritime patrol versions A and B on offer. Economical cruising speed is 332km/h and range (Do228-200) is 1,150km. *Country of origin:* West Germany. *Silhouette:* -200. *Picture:* 228 Maritime version A.

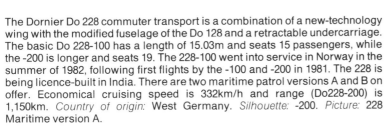

Aérospatiale N 262/Frégate *Confusion:* Turbolet

Power: 2 × Bastan turboprops   *Span:* 21.9m   *Length:* 19.28m

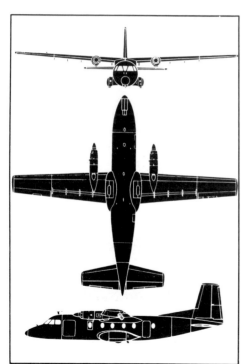

Descended from an earlier unpressurised transport, the Nord N 260 Super Broussard, the pressurised N 262/Frégate short-range airliner is used by civil and military operators. First flown in 1962, the 262 has been produced in four series: A, B, C and D. Only four Bs were built and the standard basic production version was the 262A. A few Series C with more powerful engines were built; military counterpart of the C is the Series D for the French Air Force. The 262C cruises at 397km/h over a range of 2,400km with 26 passengers. *Country of origin:* France. *Silhouette and picture:* Series C.

Grumman S-2/C-1 Tracker/Trader

Power: 2 × R-1820 piston engines   *Span* (S-2D to -2G): 22.13m   *Length:* 13.26m

Used on land or aboard aircraft carriers, the Grumman S-2 Tracker is an anti-submarine aircraft. A carrier on-board delivery version, the C-1 Trader, can seat nine passengers. The Tracker first flew in 1953 and carries homing torpedoes, depth bombs and depth charges internally and torpedoes, rockets or bombs under the wings. Various versions were built, from the S-2A to the -2G, and over 1,000 of the type were produced. With a normal crew of four, the Tracker has a maximum speed of 426km/h and nine hours' endurance. Non-US Tracker operators include Canada, Argentina, Turkey, Peru, Brazil, South Korea and Taiwan. *Country of origin:* USA. *Silhouette:* S-2D. *Picture:* S-2A.

Fokker F27 Friendship

Confusion: 'Coke', 'Curl', Tracker

Power: 2 × Dart turboprops   *Span:* 29m   *Length* (Mk 500): 25.06m

Orders for the F27 Friendship short/medium-haul airliner totalled over 786 when production ceased in 1986. First flown in 1955 the Friendship was built under licence in the USA as the FH-227. Variants are the Mks 100, 200, 300, 400/600 and the long-fuselage, 52-passenger 500. The Mk 500 also has a large side freight door. Friendships are in use as military transports. Military versions include F27 Maritime and Maritime Enforcer. The Mk 500 has a cruising speed of 480km/h and a range of 1,741km. *Country of origin:* Netherlands. *Main silhouette:* Mk 500; *upper side view:* FH-227. *Picture:* F27 Maritime.

Power: **2 × PW125 turboprops**   *Span:* 29m   *Length:* 25.24m

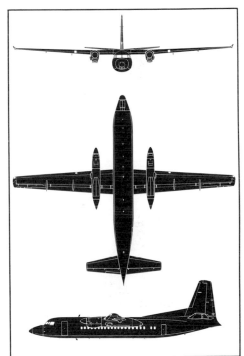

The follow-on to the widely used Fokker F-27 Friendship, the Fokker 50 has new engines in revised nacelles, six blade propellers and more cabin windows. Widespread use is made of composite materials. Seating up to 58 passengers in a high density layout, the Fokker 50 cruises at 522km/h and has a normal range of 1,163km. The Fokker 50 first flew in December 1985 and as of May 1989, 98 had been ordered. *Country of origin:* Netherlands.

Antonov An-24 'Coke'/An-26 'Curl'

Confusion: Friendship, 'Clank', 'Cline'

Power: 2 × Ivchenko turboprops   *Span:* 29.2m   *Length:* 23.53m

Over 1,000 Antonov An-24 transports, code-named 'Coke', were built after the type first flew in 1960. Initially fitted with 44 seats, the aircraft was developed to accommodate 50 passengers as the -24V. A subsequent variant was fitted with higher-powered engines and modified wing. An under-fuselage rear door is fitted to the -24T, while the -24RT and -24RV each have an auxiliary turbojet in the starboard nacelle. The An-26 'Curl' is a development with an upswept rear fuselage and enlarged rear loading ramp. The An-24 is in production in China as the Y7-100 with Western avionic equipment. *Country of origin:* USSR. *Silhouette:* An-26. *Picture:* Y7-100.

Power: 2 × Ivchenko turboprops   *Span:* 29.2m   *Length:* 24.26m

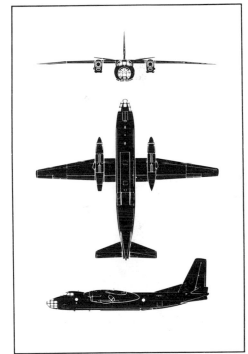

Designed for aerial survey, the An-30, code-named 'Clank', is a direct development of the An-24 'Coke'. It is fitted with a longer glazed nose and a raised flight deck. Cameras are mounted in the fuselage, which has its own darkroom, and a computer controls the flight profile. The An-30 has fewer windows than the passenger versions of the An-24. An auxiliary turbojet is mounted in the starboard nacelle. This is used for engine starting, and for take-off, climb and cruise power in the event of main engine failure. The An-30's primary role is aerial photography for map-making, for which it is equipped with large survey cameras mounted above four apertures in the cabin floor. The openings are covered by doors which can be opened under remote control from the crew photographer's station. *Country of origin:* USSR.

Antonov An-32 'Cline'

Confusion: 'Coke', 'Curl', 'Clank', Friendship

Power: 2 × Ivchenko turboprops   *Span:* 29.2m   *Length:* 23.8m

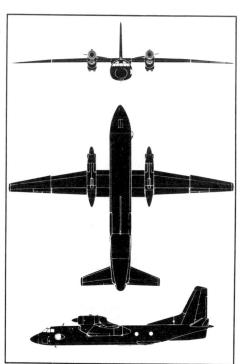

A major increase in power is the main feature of the An-32 'Cline', which is a direct development of the An-24 'Coke'. Main differences are bigger engine nacelles mounted over the wings and a greatly enlarged ventral fin. Higher power gives the 'Cline' better airfield performance, particularly at high altitudes and temperatures. Payloads include 39 passengers, 30 parachutists or 24 stretchers. A rear loading door is fitted. Range is 2,200km. Operating from airfields 4,000-4,500m above sea level, the An-32 can carry 3,000kg of freight for 1,100km. *Country of origin:* USSR.

Power: 2 × R-2000 piston engines   *Span:* 29.15m   *Length:* 22.13m

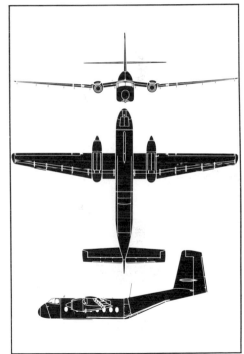

A short take-off and landing (STOL) utility transport, the DHC-4A Caribou first flew in 1958. The civil variant accommodates 30 passengers, while in military service 32 troops, 3,050kg of freight, two jeeps or stretchers can be carried. Under the upswept rear fuselage there is a large single loading door. In Canadian Forces service the Caribou is known as the CC-108. The US Army ordered 159 CV-2 Caribous. These aircraft were later taken over by the USAF, which designated them C-7A. Range is 2,103km and cruising speed 293km/h. *Country of origin:* Canada.

DHC DHC-5 Buffalo/Transporter *Confusion:* Caribou

Power: 2 × CT64 turboprops   *Span:* 29.26m   *Length:* 24.08m

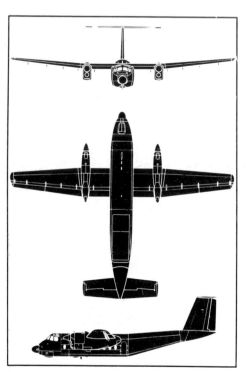

Originally called the DHC-5 Caribou II, the Stol turboprop development of the Caribou was renamed Buffalo. A military transport, the Buffalo carries 41 troops, jeeps, stretchers or freight loaded through an underfuselage rear door. The Caribou first flew in 1964. In USAF service it is known as the C-8A, and as the CC-115 with the Canadian Forces. Specially equipped Buffalos are used for maritime patrol. Range is 1,112km at a maximum cruising speed of 420km/h. The 44-seat civil passenger/cargo version of the DHC-5D Buffalo is the DHC-5E Transporter. *Country of origin:* Canada.

Power: 2 × PW120 turboprops   *Span:* 25.89m   *Length:* 22.25m

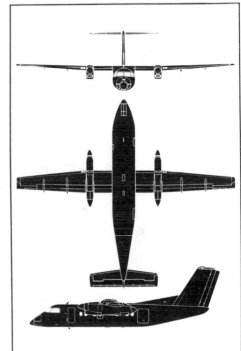

A 30/40-seat short-haul transport, the de Havilland Canada Dash 8 first flew in June 1983. There are two basic versions, the Commuter and the Corporate, the latter carrying 17 passengers over longer ranges. The Corporate has a maximum cruising speed of 500km/h and can fly four 185km stages without refuelling. First flown in May 1987, the Dash 8 Series 300 has fuselage extended by 3.43m and seats up to 56 people. Military versions of the Series 100 include the CC-142 and CT-142 for Canada and E-9A for surveillance/communications with the USAF. By March 1989 297 orders and options had been placed for the Dash 8 programme. *Country of origin:* Canada. *Main silhouette:* Dash 8; *Lower side view:* Dash 8 Series 300. *Picture:* E-9A.

Fairchild C-123 Provider

Confusion: G222, Transall

Power: 2 × R-8000 piston engines and 2 × J44 turbojets   *Span:* 33.53m   *Length:* 23.93m

A tactical assault transport, the C-123 Provider can carry up to 60 equipped troops, stretchers or cargo. It was first flown in 1949 and production totalled 307. Originally produced as C-123Bs, 183 Providers were later converted to C-123K standard with underwing auxiliary jets. A further 10 were fitted with wingtip jets as the C-123H. Equipped with spraying gear for mosquito control, the Provider becomes the UC-123B. Underwing fuel tanks are standard, while some Providers have nose radomes. C-123s were supplied to air forces in the Far East and South America. *Country of origin:* USA. *Silhouette:* C-123K. *Picture:* C-123H.

Power: 2 × T64 turboprops   *Span:* 28.7m   *Length:* 22.7m

A general-purpose pressurised military transport, the G222 is in production for the Italian Air Force and export customers. Capable of operating from rough strips, the 222 can carry 44 equipped troops, 36 stretchers or freight loaded through a large underside rear door. The prototype 222 first flew in 1970 and the type can also be used for firefighting (G222SAA), radio/radar calibration (G222RM), maritime surveillance and electronic warfare (G222VS). With Rolls-Royce Tyne turboprops the type is known as the G222T. Range of the G222 is 700km and continuous cruising speed is 360km/h. Eighty-eight 222s have been ordered. *Country of origin:* Italy. *Silhouette:* G222.*Picture:* G222T.

Transporter Allianz C-160 Transall *Confusion:* G222, Provider, Aviocar

Power: 2 × Tyne turboprops   *Span:* 40m   *Length:* 32.4m

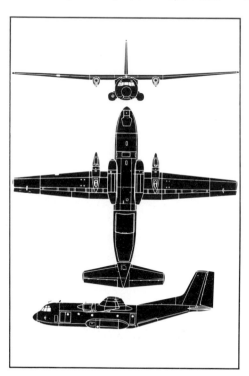

A joint Franco-German development, the C-160 Transall is a general-purpose military transport. It can carry up to 93 troops, freight or vehicles. A total of 179 were produced before production ceased in 1972. The line was later re-opened and production resumed with the Transall 'Second Series' until it closed in 1985. The aircraft of the new batch have updated electronics, increased maximum take-off weight, and a reinforced wing embodying an optional fuel tank. As well as France and Germany, South Africa and Turkey also use the Transall. The aircraft can operate from semi-prepared surfaces. Cruising speed is 492km/h and range 1,175km. *Country of origin:* France/Germany.

Power: 2 × CT7 turboprops *Span:* 25.81m *Length:* 21.35m

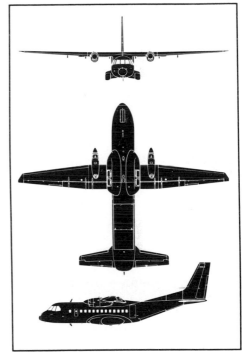

A joint venture by CASA of Spain and PT Nurtanio of Indonesia, the Airtech CN-235 is a twin turboprop commuter transport which first flew in November 1983. Seating up to 44 passengers, the CN-235 can fly four 185km stage lengths on short haul routes before needing to refuel. Maximum cruising speed is 452km/h. Production lines have been set up in both countries and, by March 1988, 114 CN-235s had been ordered. Initial production aircraft were designated Series 10 while, with up-rated engines, later production machines are called Series 100. *Countries of origin:* Spain and Indonesia

Avions de Transport Régional ATR42/ATR72

Confusion: Transall, Buffalo

Power: 2 × PW 120 turboprops   *Span:* 24.57m   *Length:* 22.67m

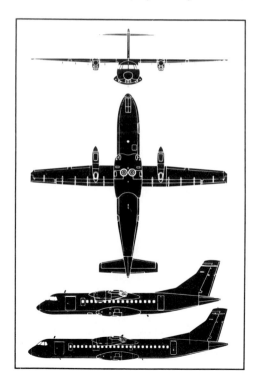

Jointly produced by Aérospatiale in France and Aeritalia in Italy, the ATR 42 commuter aircraft made its first flight in August 1984. Seating 40/50 passengers, the ATR 42 cruises at 463km/h over ranges up to 1,760km. Variants include ATR 42F freighter, ATM 42L military freighter and SAR 42 for search and rescue. Due to commence delivery in mid 1989 was a stretched fuselage version, designated ATR72. Seating up to 74 passengers and with uprated engines the ATR72 has a span of 27.05m. By March 1989 the ATR42/ATR72 order book stood at 380. *Countries of origin:* France/Italy. *Main silhouette:* ATR 42; *lower side view:* ATR 72. *Picture:* ATR 42.

Power: 2 × M602 turboprops   *Span:* 25.6m   *Length:* 21.4m

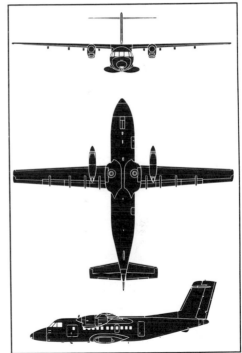

First flown in December 1988, the L610 is a larger follow-on to the successful L410 commuter aircraft. Seating 40, the L610 is powered by two new engines made by the Motorlet National Corporation. Compared with the L410, the L610 has increased span and length, circular section fuselage and tailplane mounted high on the tall fin. Deliveries are due to begin in 1991, with Aeroflot expected to acquire some 600. The L610 cruises at 400km/h and has a range of 870km. *Country of origin:* Czechoslovakia.

 **CASA C-212 Aviocar** *Confusion:* Transall

Power: 2 × TPE331 turboprops   *Span:* 19m   *Length:* 15.16m

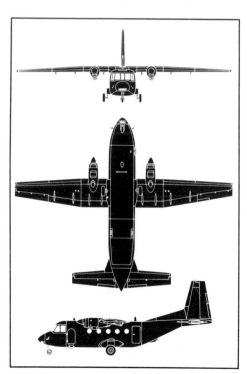

Used for both military and civil purposes, the Aviocar is a Stol utility transport, trainer and survey aircraft. More than 400 have been sold, and there are production centres in Spain and Indonesia. The C-212-5 Series 100, the first production version, was superseded on the line by the Series 200 with more powerful engines and greater weight. The latest production version, with extended and other improvements, is the Series 300. In civil transport form the Aviocar Series 200 seats up to 26 passengers, cruises at 347km/h and has a range of 370km with full payload. An anti-submarine and maritime patrol version with a nose radome, has been developed. *Country of origin:* Spain. *Silhouette:* Series 200. *Picture:* Series 100.

Power: 2 × **PT6A turboprops** *Span:* 17.23m *Length:* 14.86m

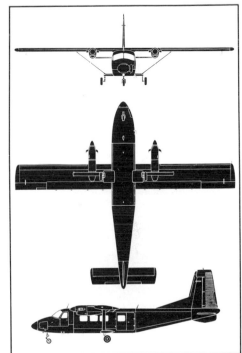

Bearing a superficial resemblance to the Pilatus Britten-Norman Islander, the Yun-12 is a short field turboprop-powered utility transport. First flown in August 1984, the type is in production and has already been exported to Sri Lanka. The Yun-12 has a crew of two and can carry up to 17 passengers or freight. In addition it can be equipped for crop spraying and other roles. Cruising speed is 328km/h, maximum take-off weight 5,300kg and take-off run 340m. *Country of origin:* China.

Pilatus Britten-Norman Islander/Defender

Confusion: Trislander, P.68 Victor, Yun-12

Power: 2 × Lycoming piston engines   *Span:* 14.94m   *Length:* (long nose): 12.02m

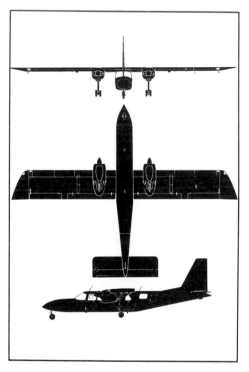

First flown in 1965, the 10-passenger BN-2 Islander light transport has sold all over the world and over 1,100 have been ordered. Licence production is going on in Romania and the Philippines. Compared with earlier versions, the BN-2A-85 Islander has a longer nose and an extra rear cabin window on each side. Military variants with underwing stores are the Defender and the Maritime Defender. Range at a cruising speed of 257km/h is 1,400km. A current version is the BN-2-T Turbine Islander, powered by two Allison turboprops. A surveillance version with a large circular nose radar is also on offer as the AEW Defender. There are also AEW/MR and ASW/ASV developments. *Country of origin:* UK. *Silhouette:* BN-2A-85. *Picture:* AEW Defender.

Pilatus Britten-Norman Trislander

Power: 3 × Lycoming piston engines   *Span:* 16.15m   *Length:* (Mk III-2): 13.33m

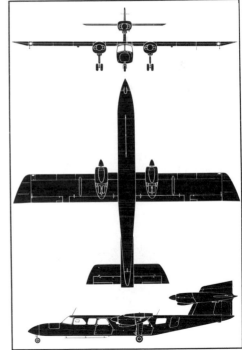

The BN-2A Trislander is derived from the twin-engined Islander and has 75 per cent of its components in common. Essential difference is the fitting of a third engine in the redesigned fin. Some Trislanders have been produced with a short nose, while the Mk III-2 has a long nose . Trislanders were produced in Belgium for a time. The aircraft carries 16/17 passengers, cruises at 282km/h and has a range with maximum payload of 257km. Over 80 Trislanders were delivered and the licence for the type is held by International Aviation Corp, where the name has been changed to Tri-Commutair. *Country of origin:* UK. *Silhouette and picture:* Mk III-2.

Partenavia P68 Victor

Confusion: Islander, Twin Otter

Power: 2 × Lycoming piston engines or 2 × Allison 250 turboprops   *Span:* 12m   *Length:* 9.35m

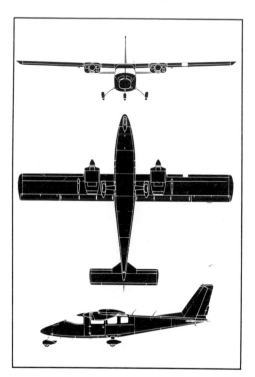

A streamlined light transport, the original P68 was first put into production in 1964. A variety of versions have emerged, including the six-passenger P68B, the P68R with retractable undercarriage, P68C, currently in production, and P68TC with turbosupercharged engines. An Italian/German development, the P68 Observer, has a transparent nose and is designed for police and observation duties. Cruising speed of the P68B is 296km/h and range 1,656km. Derivatives of the P68 are the larger AP68TP-300 Spartacus and AP68TP-600 Viator, the latter with a retractable undercarriage. *Country of origin:* Italy. *Silhouette:* P68C. *Picture:* P68R.

Power: 2 × Lycoming piston engines   *Span:* 13.8m   *Length:* 9m

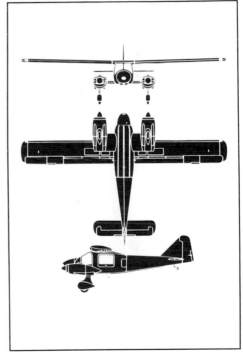

A twin-engined version of the Do 27, the Do 28 seven-passenger light transport retains the basic wings and fuselage of the single-engined type. Some 120 Do 28s were built, including military aircraft and six floatplanes. First flight was in 1959. The nose was redesigned, the tailplane enlarged and wingspan increased to produce the Do 28B1. The B2 has supercharged engines. Maximum cruising speed of the B1 is 274km/h and range 1,235km. *Country of origin:* West Germany. *Silhouette:* Do 28B1. *Picture:* Do 28B2.

Dornier Do 28D/Do 128 Skyservant

Confusion: Do 28B

Power: 2 × Lycoming piston engines or 2 × PT6A turboprops   *Span:* 15.55m   *Length:* 11.41m

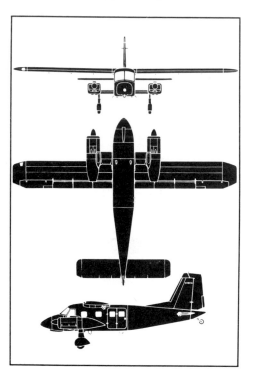

Retaining the same layout as the Do 28 but virtually a complete redesign, the Do 28D Skyservant first flew in 1966. A Stol utility transport, the Skyservant seats up to 14 passengers or can carry freight. Six air forces have bought Skyservants, the largest operator being the Luftwaffe which still has 55. The turboprop-powered Do 128-6 with PT6A engines, followed the end of Do 128-2 Skyservant production. *Country of origin:* West Germany. *Silhouette:* Do 28D2. *Picture:* Do 128-6.

Power: 2 × Allison turboprops   *Span:* 16.46m   *Length:* 14.35m

A twin-turboprop high-wing utility transport and feederliner, the ASTA (formerly GAF) Nomad first flew in 1971. The prototype was known as the N2, while the initial production variant was the 17-passenger N22B. Patrol and surveillance versions of the N22B are known as the Search Master B and L. For general-purpose military work the N22B is known as the Missionmaster. An uprated version is the N22C. The N24 has increased fuselage length and seating for 19 passengers. The N24 Nomad has a range of 1,585km and a cruising speed of 269km/h. A total of 170 Nomads were built. *Country of origin:* Australia. *Silhouette:* N22B. *Picture:* Search Master.

DHC DHC-6 Twin Otter   *Confusion:* Nomad, 'Clod', 'Cash'

Power: 2 × PT6 turboprops   *Span:* 19.81m   *Length:* 15.77m

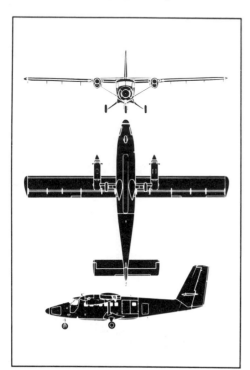

A general-purpose civil and military transport, the DHC-6 Twin Otter can operate from short, rough fields and accommodate up to 20 passengers or freight. First flown in 1965, the Twin Otter has been built in three forms, the 100, 200 and 300 series. Service designations are CC-138 (Canadian Forces) and UV-18A (US Army). Maximum cruising speed is 337km/h, at which the range is 1,775km. The floatplane version has a short nose and small additional fins above and below the tailplane. A Twin Otter Series 300 supplied to China for survey work has a nose probe and wingtip pods. *Country of origin:* Canada. *Silhouette:* Series 200. *Picture:* Series 300.

Power: 2 × Ivchenko piston engines   *Span:* 21.99m   *Length:* 11.44m

A rough-field Stol utility transport, the An-14, code-named 'Clod', was produced in quantity, mainly for Aeroflot. With a one-man crew, the 'Clod' has seating for seven passengers. The prototype flew in 1958 but production was delayed for a major redesign to the present standard, with a high-aspect-ratio wing and large endplate fins. In the Soviet Union the 'Clod' is known as the *Pchelka* ('Little Bee'). Clamshell rear doors simplify freight loading. Cruising speed is 180km/h. *Country of origin:* USSR.

PZL Mielec (Antonov) An-28 'Cash'

Confusion: 'Clod', Skyvan, Shorts 330, 360

Power: 2 × TVD-10B turboprops   *Span:* 22.06m   *Length:* 13.1m

Originally developed by the Antonov Bureau in the Soviet Union as a turboprop version of the An-14 'Clod', the An-28 'Cash' is now produced in Poland by WSK-PZL Mielec. The 'Cash' has a high aspect-ratio braced wing, sponson-mounted undercarriage and rear under-fuselage doors. It is intended for a variety of rôles including flying training, carriage of passengers (17), mail and freight, fire-fighting, geological survey and agricultural work. Initial deliveries from the Polish line began in 1984. The An-28 cruises at 337km/h over ranges up to 350km. *Countries of origin:* USSR/Poland.

Power: 2 × TPE331 turboprops   *Span:* 19.79m   *Length:* 12.21m

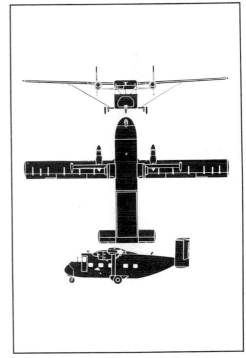

A rugged Stol general-purpose transport, the Skyvan originally flew in 1963 with piston engines. It was then fitted with French Astazou turboprops and finally with Garrett AiResearch turboprops. Some 150 Skyvans were built for military, paramilitary and civil use. The Skyliner feederliner can carry up to 19 passengers, while freighter versions can lift 2,085kg of cargo or vehicles. Freight is loaded through an underside rear door. Skyvan Series 3 has a range of 1,115km and a maximum cruising speed of 314km/h. *Country of origin:* UK. *Silhouette and picture:* Series 3.

Shorts 330/Sherpa/C-23A

Confusion: Skyvan, Shorts 360, 'Cash', Skyservant

Power: 2 × PT6A turboprops   *Span:* 22.76m   *Length:* 17.69m

A 30-seat feederliner and utility transport, the Shorts 330 is similar in basic configuration to the smaller Skyvan. The fuselage is rectangular in cross-section and the wing has a high aspect ratio. Unlike the Skyvan, the 330 has a retractable undercarriage. Over 170 330s have been ordered, including the standard 330-200 passenger version, the 330-UTT military utility transport and the Sherpa freighter with rear door, 18 of which were ordered by the USAF as the C-23A. Cruising speed of the 330-200 is 296km/h and range with maximum fuel, 1,695km. *Country of origin:* UK. *Silhouette:* 330. *Picture:* C-23A Sherpa.

Power: 2 × **PT6A** turboprops   *Span:* **22.75m**   *Length:* **21.49m**

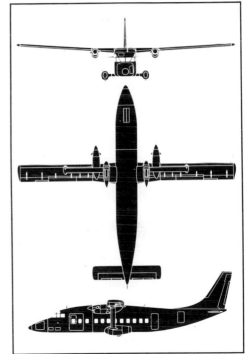

Retaining the high-aspect-ratio wing of the Shorts 330, the 360 differs from the former in having a 0.91m extension in the forward fuselage and a redesigned rear fuselage with a single swept fin. Flown for the first time in 1981, the 360 is in production at Belfast alongside the 330. A short-range commuter transport, the 360 carries 36 passengers and cruises at 391km/h, and has a maximum range of 1,054km. Latest version is the 360-300 with up-rated engines, and six blade propellers. Deliveries began in the last quarter of 1982 and by March 1989 deliveries totalled 149. *Country of origin:* UK. *Silhouette:* 360. *Picture:* 360-300.

Grumman E-2 Hawkeye

Confusion: —

Power: 2 × T56 turboprops   *Span:* 24.56m   *Length:* 17.55m

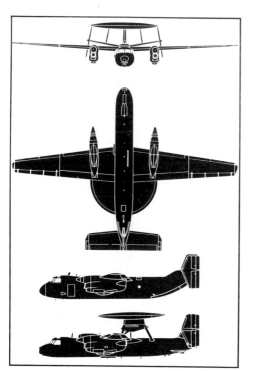

The first US Navy turboprop-powered type to enter service, the E-2 Hawkeye is a carrierborne early-warning aircraft. Virtually an airborne radar station, the Hawkeye carries a massive oval radome mounted above the fuselage and houses complex electronic gear operated by a five-man crew. The differences between the various Hawkeye versions (E-2A, B and C) lie in the radar and internal equipment. The transport version of Hawkeye is the C-2A Greyhound, which can accommodate 39 passengers, stretchers or cargo. The aircraft has a range of 2,660km. *Country of origin:* USA. *Main silhouette:* E-2C; *upper side view:* C-2A. *Picture:* E-2C.

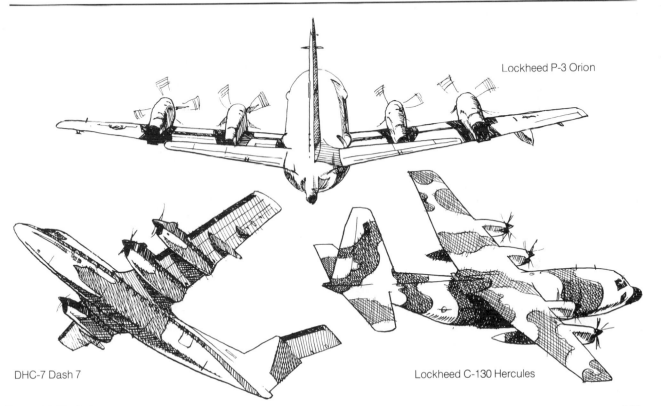

Lockheed P-3 Orion

DHC-7 Dash 7

Lockheed C-130 Hercules

de Havilland Heron *Confusion:* Viscount

Power: 4 × Gipsy Queen piston engines   *Span:* 21.8m   *Length:* 14.8m

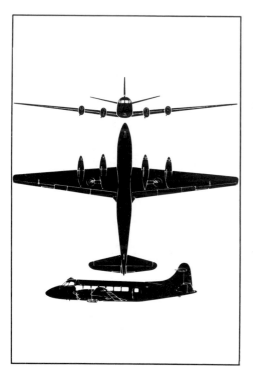

A scaled-up, four-engined Dove, the Heron first flew in 1950 and altogether 150 were built. As a short-range light transport the Heron seats 14/17, cruises at 362km/h and has a range of 1,900km. Herons are used as executive aircraft and for military purposes. Re-engined conversions, include the Shin Meiwa DH 114TAW, the Saunders ST-27 and the Riley Turbo Skyliner. Powered by four 290 hp Lycoming engines with Rajay turbo-superchargers, the Turbo Skyliner has a maximum level speed at 3,660m of 495km/h. A total of 18 Riley Turbo Skyliners were converted from Heron Series 2X, 2A and 2DA aircraft. *Country of origin:* UK. *Silhouette:* Heron. *Picture:* Riley Turbo Skyliner.

Power: 4 × **Dart turboprops**   *Span:* (700): 28.56m   *Length:* (700): 25.04m

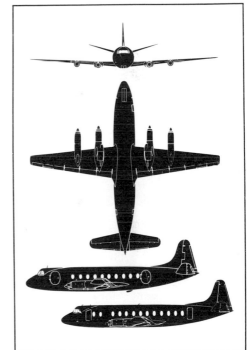

The world's first turboprop transport to go into service, the Viscount short-range airliner first flew in 1948 and 441 were built. The first production version was the Series 700 with up to 59 passengers. This was followed by the stretched 800 series with seating for up to 65 passengers. The Viscount remains in service in many countries. Several 700s and 800 were built or converted for executive use. Scottish Aviation designed a convertible passenger/cargo interior for Series 800 Viscounts. Nine pallets can be loaded and maximum cargo weight is 6,759kg. The 700 cruises at 502km/h and has a range of 2,815km. *Country of origin:* UK. *Main silhouette:* Viscount Series 800; *upper side view:* Viscount Series 700. *Picture:* Viscount Series 700.

Ilyushin Il-18/Il-20 'Coot'

Confusion: 'May', Orion, Electra

Power: 4 × Ivchenko turboprops   *Span:* 37.4m   *Length:* 35.9m

Over 700 Il-18 'Coot' medium-range airliners were built and the type was exported to a number of countries. It first came into service in 1959 and was known in Russia as the 'Moskva'. Several versions were produced, including the Il-18V with 89/100 passengers and the Il-18D with 122 passengers. The Il-18D cruises at 625km/h over ranges of up to 6,500km. A military development is the electronic surveillance Il-20 'Coot A', with forward fuselage side bulges, a large underfuselage pannier and dorsal aerial fairings. An airborne command post variant code-named 'Coot B', is believed to house sideways-looking radar. *Country of origin:* USSR. *Silhouette:* Il-18. *Picture:* Il-20 'Coot A'.

Power: 4 × Ivchenko turboprops *Span:* 37.4m *Length:* 39.92m

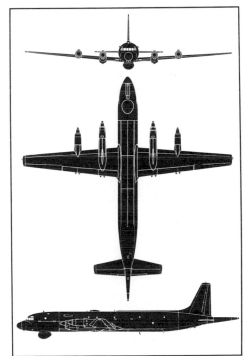

Like the American Orion, the Il-38, code-named 'May', is a civil airliner developed for maritime reconnaissance and anti-submarine warfare. The 'May' marries the wings and tail assembly of 'Coot' to a new fuselage with a bulbous radome under the forward section and a magnetic anomaly detector (MAD) fairing extending from the tail. 'May' has a cruising speed of 595km/h and a range of 7,240km. Armament includes homing torpedoes and depth charges. About 60 Il-38s are in Soviet service and three have been exported to India. *Country of origin:* USSR.

Lockheed L-188 Electra

Confusion: Orion, 'Coot', 'May'

Power: 4 × Allison 501 turboprops   *Span:* 30.18m   *Length:* 31.81m

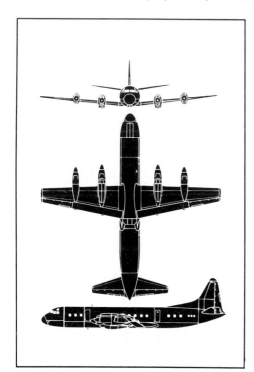

A medium-range four-turboprop airliner, the Lockheed Electra first flew in 1957. Early in its career the Electra suffered two disastrous crashes, the result of metal fatigue. A number of modifications were required by the airworthiness authorities, including strengthening of the wing and engine nacelle structure, and the provision of thicker wing skins. Two versions of the Electra were built: the L-188A and the longer-range L-188C. A total of 168 Electras were completed. The Orion maritime reconnaissance aircraft was evolved from the Electra. The L-188C seats 98 passengers, cruises at 652km/h and has a range of 4,458km. *Country of origin:* USA. *Silhouette:* L-188C. *Picture:* L-188A.

Power: 4 × T.56 turboprops   *Span:* 30.37m   *Length:* 35.61m

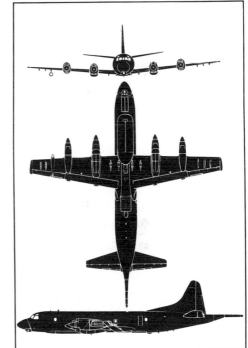

A development of the civil Lockheed Electra, the P-3 Orion long-range anti-submarine aircraft first flew in 1958. Carrying a crew of ten, the Orion is equipped with sophisticated search equipment, including the distinctive tail-mounted magnetic anomaly detector (MAD) probe. Armament includes torpedoes and depth charges. A number of versions of the Orion have been built, including the CP-140 for Canada and EP-3 ELINT aircraft. The avionics of the Orion have been continuously up-dated. Over 620 Orions have been delivered. The P-3C Orion has a four-engine loiter endurance of 12.3h and patrol speed is 381km/h. An AEW version of the Orion is the P-3 Sentinel, with a large circular radome mounted above the rear fuselage. *Country of origin:* USA. *Silhouette and picture:* P-3C.

Douglas DC-4/C-54 Skymaster

Confusion: DC-6, DC-7

Power: 4 × R-2000 piston engines   *Span:* 35.8m   *Length:* 28.6m

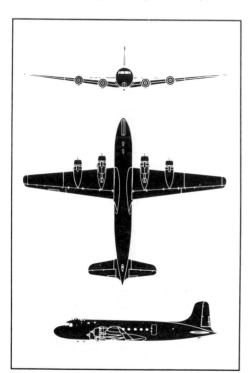

First flown as long ago as 1938, the DC-4 was widely used in military as well as civil forms. The Air Force variant was known as the C-54 Skymaster and the naval version as the R-5D. As an airliner the DC-4 seats 44 passengers, cruises at 363km/h and has a range of 2,480km. Military versions have a large freight door on the port side. Many Skymasters later found their way onto the civil market. About 30 Skymasters remain in military service round the world. *Country of origin:* USA.

Confusion: DC-4, DC-7 **Douglas DC-6**

Power: 4 × R-2800 piston engines   *Span:* 13.58m   *Length:* 30.66m

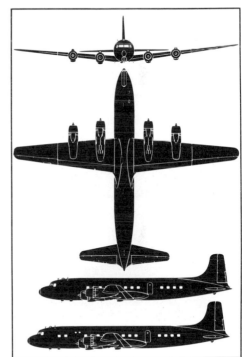

A stretched, pressurised version of the DC-4, the DC-6 first flew in 1946 and 176 were built. Some air forces also acquired DC-6s. The DC-6A was a freighter with a new fuselage, while the DC-6B was a 54-passenger airliner version of the DC-6A. The military DC-6A was designated C-118A. Cruising speed of the DC-6 is 501km/h and range 6,112km. The C-118 Liftmaster, still widely used by air forces around the world, can carry 76 fully equipped troops, 60 stretchers or 12,247kg of cargo. Large freight loading doors are fitted fore and aft of the wing. With a maximum level speed at 5,520m of 579km/h, the C-118 has a normal range of 6,212km. A total of 167 Liftmasters were built. *Country of origin:* USA. *Main silhouette:* DC-6B; *upper side view:* DC-6. *Picture:* C-118.

351

Douglas DC-7B/C

Confusion: DC-4, DC-6, Britannia

Power: 4 × R-3350 piston engines   *Span:* (-7C): 38.86m   *Length:* 34.21m

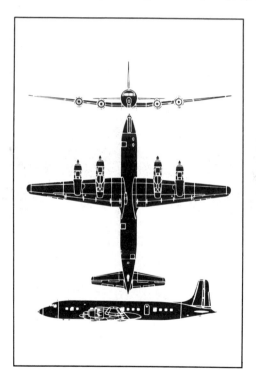

A stretch of the DC-6B fuselage produced the DC-7, seating up to 95 coach-class passengers. A change to Wright turbo-compound engines raised weight and improved performance. More fuel was provided on the -7B, which had intercontinental range. The last of the DC-4/DC-7 line was the -7C with increased wingspan and a longer fuselage. Some 120 DC-7Cs were built and a number were converted to DC-7CF freighters. The DC-7C cruises at 499km/h for up to 6,595km. Production of the DC-7 series ran to 120 DC-7s, 96 DC-7Bs and 120 DC-7Cs; these totals include some aircraft converted to DC-7F freighters. *Country of origin:* USA. *Silhouette and picture:* DC-7C.

Power: 4 × Griffon piston engines   *Span:* 36.52m   *Length:* 26.62m

Forty years after it first flew, the venerable Shackleton is still in RAF front-line service. Originally produced for maritime reconnaissance, the Shackleton was later converted for airborne early warning as the AEW2. Fitted with a large radome under the forward fuselage, the AEW2 carries a crew of 12 and can remain on station for up to 15h. The type is due to be superseded in the RAF by the Boeing E-3 Sentry from 1991 onwards. *Country of origin:* UK. *Silhouette and picture:* AEW2.

353

Power: 4 × R4360 piston engines or 4 × Allison 501 turboprops *Span:* 47.62m *Length:* 43.84m

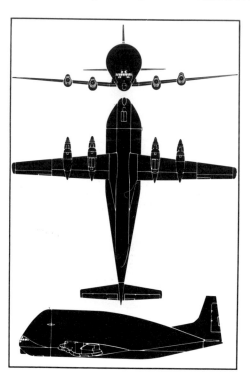

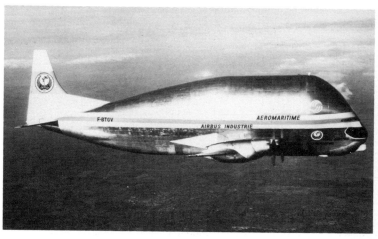

Retaining the wings and engines of the Boeing Stratocruiser airliner, the Aero Spacelines Guppy has a vast fuselage originally designed to accommodate space launcher stages. Subsequently the type has been used to airlift other specialised freight, including wings for the Airbus A300. In 1980 Airbus Industrie acquired production rights to the design. Range with maximum payload is 813km and maximum cruising speed is 463km/h. The normal flight crew is four. Variants are known as Pregnant Guppy, Super Guppy and Mini-Guppy. *Country of origin:* USA. *Silhouette and picture:* Guppy.

Power: 4 × T56 turboprops   *Span:* 40.41m   *Length:* 29.78m

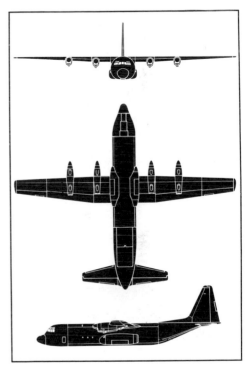

One of the most successful military transports built in the West, the C-130 Hercules first flew in 1954 and has been in production since 1955. Over 1,850 have been built for operators in 55 nations. There are many versions, including the C-130D for Arctic work, WC-130B and -130E for weather reconnaissance, AC-130E gunship, KC-130F Marine Corps tanker, C-130F for the US Navy, -130H and -130N for rescue, -130P for helicopter refuelling , EC-130Q for Navy communications and C-130K for the RAF. Twin under-body strakes are being fitted to many Hercules and 30 RAF K-130s have a 4.57m fuselage-extension, designated Hercules C Mk 3. A civil version is designated L-100. The C-130H, which carries 92 troops or freight, cruises at 547km/h and has a range with maximum payload of 3,943km. *Country of origin:* USA. *Silhouette:* L-100. *Picture:* KC-130T.

355

Antonov An-12 'Cub'

Confusion: Hercules, Belfast

Power: 4 × Ivchenko turboprops   *Span:* 38m   *Length:* 37m

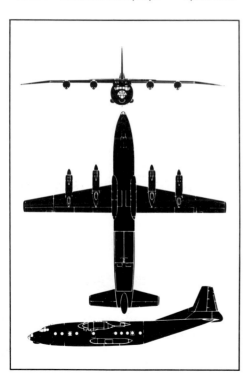

Produced in very large numbers, the An-12, code-named 'Cub', is used in both military and civil forms. The 'Cub' has a crew of five and can carry 100 paratroops, vehicles, guns or missiles, loaded through the large rear ramp. 'Cubs' serving with the Soviet and other air forces have a twin 23mm gun turret in the tail. This is replaced by a fairing on the civil version, which is known as the An-12V. Electronic intelligence and electronic countermeasures versions are coded 'Cub B' and 'Cub C' and 'Cub D'. The 'Cub' cruises at 550km/h and has a range of 3,400km. *Country of origin:* USSR. *Silhouette:* An-12V; *Picture:* 'Cub A'.

Power: 4 × PT6 turboprops   *Span.* 28.35m   *Length:* 24.58m

Capable of very short take-offs and landings, the DHC-7 Dash 7 seats 50 passengers or carries a mixture of passengers and freight. First flown in 1975, over 100 Dash 7s are in service. The Dash 7 operates over short routes and, in addition to its good airfield performance, is very quiet in operation. Cruising speed is 379km/h and range is 1,120km. Initial version was the Series 100 and this was followed by the longer range Series 150. All-cargo versions are designated Series 101 and 151. *Country of origin:* Canada. *Silhouette and picture:* Dash 7.

Antonov An-22 'Cock'

Confusion: 'Cub'

Power: 4 × Kuznetsov turboprops   *Span:* 64.4m   *Length:* 57.8m

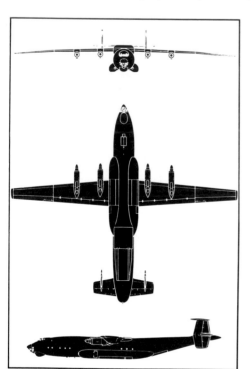

The Antonov An-22 Antheus, code-named 'Cock', is a heavy freighter for military and civil use. A crew of five/six, 28/29 passengers and a wide variety of vehicle, gun, missile and freight loads can be carried. The first 'Cock' flew in 1965 and the type went into service with the Soviet Air Force in 1967. Some 50 of the type remain in service. An unusual feature is the fitting of two radomes in the nose. The four turboprop engines drive large contra-rotating propellers. 'Cock' has a range of 10,950km and cruises at 679km/h. The An-22 set a number of world records, including no fewer than 14 payload-to-height marks on one occasion in 1967. *Country of origin:* USSR.

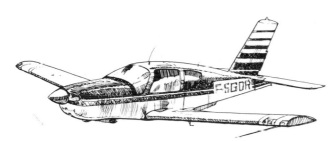

SOCATA TB 20 Trinidad

SIAI-Marchetti SF260C

Beech Bonanza V35

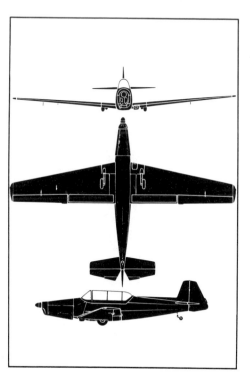

Zlin 526/726

Confusion: Meta-Sokol

Power: 1 × Avia piston engine *Span:* 9.88m *Length:* 7.98m

The Zlin Trener, Trener-master and Akrobat series has been in production in Czechoslovakia since 1947; more than 1,400 aircraft have been built. First of the series to feature a retractable undercarriage was the Z 326 of 1957. The single-seat competition aircraft were designated Akrobats and scored a number of successes in international aerobatics contests. Various engines have been used but all models have the distinctive swept wing leading edge and large, angular rudder. Last of the series is the Z 726, which first flew in 1973. Generally similar to the Z 526F, it has a slightly shorter wing and metal-covered rudder and elevators. Production ended in 1977 after 32 had been built. Maximum cruising speed of the 526F is 210km/h. *Country of origin:* Czechoslovakia. *Silhouette:* Zlin 526. *Picture:* Zlin 526A Akrobat.

CZAAL L-40 Meta-Sokol

Power: 1 × Walter Minor piston engine   *Span:* 10.06m   *Length:* 7.54m

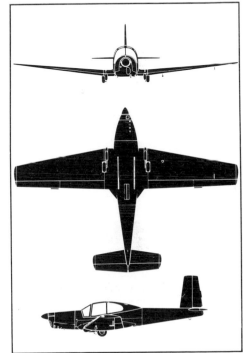

Over 200 L-40 Meta-Sokols were built between 1954 and 1961, most of which were exported. A number remain in service in Europe. On the ground the type's most easily identified feature is its unique reverse tricycle undercarriage. The main wheels are half-concealed when retracted. Also distinctive is the tall, almost parallel fin and rudder. Cruising speed is 220km/h, maximum range 1,107km, maximum rate of climb at sea level is 192m/min, and service ceiling 4,500m. *Country of origin:* Czechoslovakia.

 **Saab Safir** *Confusion:* Neiva Universal

Power: 1 × Lycoming piston engine   *Span:* 10.6m   *Length:* 8.03m

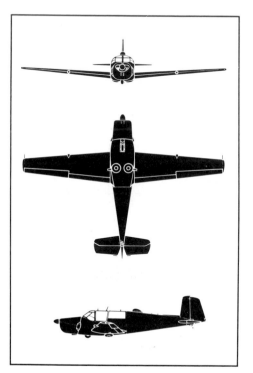

First flown in 1945, the Saab-91 Safir was built in four main versions: the Saab-91A, powered by a 145hp de Havilland Gipsy Major 10; the Saab-91B, powered by a 190hp Lycoming; the Saab-91C, widely used for private and light commercial work and as an air force trainer; and the Saab-91D, generally similar to the -91C, but powered by a 180hp Lycoming driving a constant-speed propeller. Military Safirs remain in service in Sweden and Austria. Others remain in private use in Europe. Maximum speed at sea level is 266km/h. *Country of origin:* Sweden. *Silhouette and picture:* Saab-91D.

Power: 1 × Lycoming piston engine   *Span:* 11m   *Length:* 8.6m

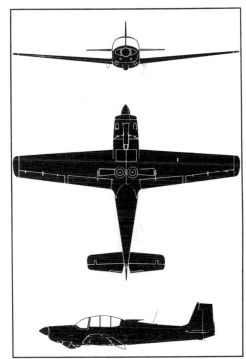

Designed as a replacement for Brazilian Air Force Fokker S-11/S-12s and North American T-6 Texans, the Neiva flew for the first time in 1966. The Brazilian Air Force designation for this tricycle-undercarriage 2/3 seater is the T-25. About 160 T-25s were delivered to the Brazilian Air Force, together with 30 of an armed version known as the AT-25. Two underwing hard-points are provided for 7.62mm machine gun pods. Some 5 T-25s are in use in Peru. Maximum cruising speed is 285km/h. *Country of origin:* Brazil.

Pilatus P-3 *Confusion:* Universal

Power: 1 × Lycoming piston engine   *Span:* 10.4m   *Length:* 8.75m

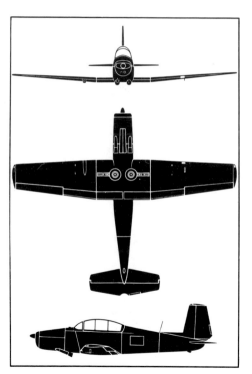

The Pilatus P-3 is an all-metal tandem two-seat trainer built primarily for the Swiss Air Force; another six aircraft went to Brazil. First flight was in 1953, with the first of 18 pre-production aircraft following the next year. Powered by a 260 hp Lycoming, the P-3 has a cruising speed of about 249km/h. Suitable for primary and advanced training, including aerobatics, night flying and instrument flying, the P-3 can be equipped for weapon training with one 7.9mm machine gun in an underwing pod, four practice bombs or two air-to-ground rockets. The P-3 will be superseded in Swiss Air Force service by the PC-7. *Country of origin:* Switzerland.

Laverda Super Falco F8L

Power: 1 × Lycoming piston engine   *Span:* 8m   *Length:* 6.5m

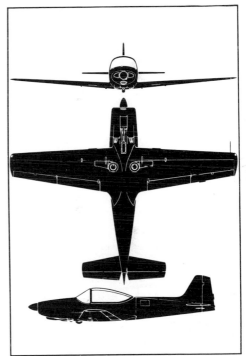

The Falco first flew in 1955, powered by a 90hp Continental. Since then four production versions have been built: F8L Series I, an initial production batch of ten built by Aviamilano in 1956 and powered by 135hp Lycomings; F8L Series II, also built by Aviamilano, with a 150hp Lycoming; F8L America, built by Aeromere for the US market; Super Falco Series IV, generally similar to the F8L America but powered by a 160hp Lycoming and built by Laverda (the former Aeromere company). An extremely clean design, the Super Falco seats two side-by-side and is fully aerobatic. Maximum cruising speed is 290km/h, and maximum sea-level climb rate 300m/min. *Country of origin:* Italy. *Silhouette and picture:* **Super Falco.**

Navion Rangemaster

Confusion: Super Falco

Power: 1 × Continental piston engine   *Span:* 10.59m   *Length:* 8.38m

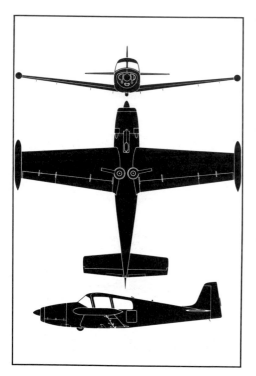

The five-seat Navion Rangemaster is a much modified development of the original Navion, produced first by North American Aviation and later, between 1948 and 1950, by Ryan, which acquired the rights in 1947. The Rangemaster retains the original Navion wing and undercarriage but is otherwise completely re-engineered. First flight was in 1960, with production commencing the next year. Fitted with wingtip fuel tanks, the type has excellent range and cruises at more than 298km/h on its 260hp Continental. A large number of Navions of all types are still active, particularly in the USA. *Country of origin:* USA.

EMBRAER EMB-312 (T-27)/Shorts S312 Tucano

Power: 1 × PT6A turboprop or 1 × TPE 331-12 turboprop   *Span:* 11.14m   *Length:* 9.86m

First flown in August 1980, the two-seat T-27 Tucano is in service with the Brazilian Air Force, which has 115 and several other air forces have Tucanos. The Tucano is designed for pilot training, but is also suitable for counter-insurgency work. Egypt has ordered 120 Tucanos. Most of the Egyptian order will be licence-built. The RAF ordered 130 of a re-engineered and developed version called Shorts S312. This is built entirely in Northern Ireland. The engine is a Garrett TPE 331 turboprop. External differences include nose, dorsal fin and cockpit hood. The S312 has also been ordered by Kuwait and, reportedly, by Kenya. *Countries of origin:* Brazil/UK. *Silhouette:* EMB-312. *Lower side view and picture:* S312.

Aérospatiale TB 30 Epsilon   *Confusion:* Tucano, PC-7

Power: 1 × Lycoming piston engine   *Span:* 7.92m   *Length:* 7.59m

The TB 30 Epsilon is the French Air Force's standard primary trainer. It entered service in 1983 and 150 are being delivered. An armed version is offered for export, fitted with four underwing hardpoints to carry a variety of stores, including two 7.62mm machine gun pods. Maximum speed of the Epsilon trainer is 380km/h and endurance 3.75h. Pilots trained on the Epsilon will progress to the Alpha Jet. The type has been exported to Portugal and Togo. *Country of origin:* France.

Power: 1 × PT6A turboprop   *Span:* 10.3m   *Length:* 8.9m

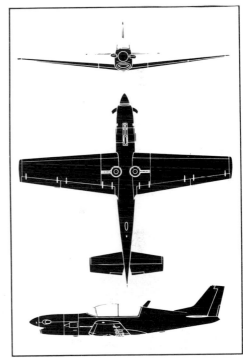

First flown on 12 June 1982, the IAR-825 has been designed as a multi-role trainer for the Romanian Air Force. Based on the piston-engined IAR-823, 60 of which were delivered to the Romanian Air Force, the -825 features a new fuselage and tail unit. In much the same class as the Brazilian Tucano, the IAR-825 is powered by a Canadian Pratt & Whitney PT6A turboprop of 680hp. *Country of origin:* Romania.

Pilatus PC-7 Turbo-Trainer

Confusion: T-34C, Tucano, Epsilon, PC-9

Power: 1 × PT6 turboprop   *Span:* 10.4m   *Length:* 9.75m

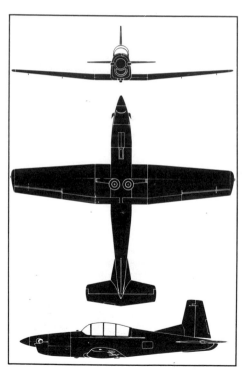

The PC-7 Turbo-Trainer is designed for basic, intermediate and aerobatic training. Over 390 have been ordered by 15 nations. Principal users are Bolivia, Iraq, Malaysia, Mexico and Switzerland, which ordered 40. All PC-7s are fitted with six underwing hardpoints as standard and up to 1,000kg of stores can be carried. At least two PC-7s have been civilian registered. Powered by a 550hp Pratt & Whitney turboprop, the PC-7 has a cruising speed of more than 300km/h. *Country of origin:* Switzerland.

Confusion: PC-7, T-34C, Tucano, Epsilon   **Pilatus PC-9**

Power: 1 × PT6 turboprop   *Span:* 10.12m   *Length:* 10.17m

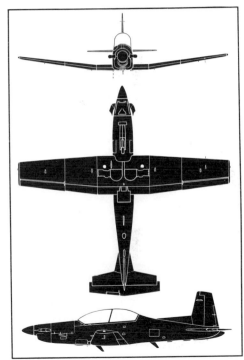

Although outwardly similar to the PC-7, the PC-9 is 90 per cent a new aircraft. Main external differences lie in the shape of the cockpit hood, longer dorsal fin and increased dihedral on the outer wing sections. First flown as a pre-production aircraft in May 1984, the PC-9 has attracted more than 126 orders, of which 48 aircraft are being licence-built in Australia for the RAAF. The PC-9 has a maximum operating speed of 593km/h and maximum range of 1,642km. *Country of origin:* Switzerland.

Beechcraft T-34 Mentor/T-34C

Confusion: PC-7, KM-2B, Tucano, Epsilon

Power: 1 × Lycoming piston engine
or 1 × P & W turboprop (T-34C)   *Span:* (T-34): 10m   *Length:* (T-34): 7.9m

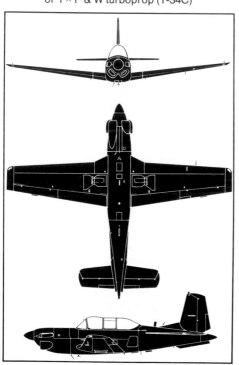

Derived from the Bonanza 33 series, more than 450 piston-engined T-34As were built for the USAF and a further 423 T-34Bs for the US Navy. The T-34C, which first flew in 1973, is extensively revised, and is powered by a 715shp Pratt & Whitney turboprop. The US Navy ordered 334 T-34Cs. Six civil versions are in service in Algeria and 119 armed T-34C-1s have been exported. *Country of origin:* USA. *Silhouette and picture:* T-34C.

Power: 1 × Lycoming piston engine　　*Span:* 10m　　*Length:* 8.04m

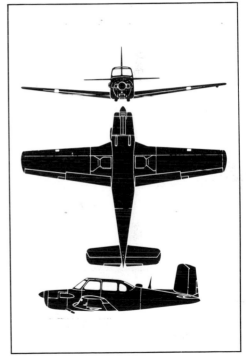

The KM-2B is a modification of the original KM-2 development of the Beechcraft Mentor, combining the airframe and powerplant of the KM-2 with the two-seat tandem cockpit of the T-34A Mentor. The KM-2B first flew in 1974 and was selected by the Japan Air Self-Defence Force (JASDF) to replace its T-34As in the primary training role the following year. The JASDF ordered 50 examples of the type, designating them T-3. The first of these made its first flight on January 17, 1978. Powered by a 340hp Lycoming flat-six engine, the KM-2B has a cruising speed of about 254km/h, maximum level speed of 367km/h and service ceiling of 8,170m. A modified version, now going into service, is the KM-2 Kai which carries the service designation T-5. *Country of origin:* Japan. *Silhouette:* KM-2. *Picture:* KM-2B.

SIAI-Marchetti S.205/S.208

Confusion: Ranger

Power: 1 × Lycoming piston engine *Span:* 10.86m *Length:* 8m

The S.205 four-seat all-metal light aircraft first flew in 1965. The type was designed from the outset to take a variety of engines, from 180 to 300hp, and there is also a version with fixed undercarriage. A later version, the S.205AC powered by a 200hp Lycoming, was ordered by the Italian Aero Club. The first of 140 were delivered in 1977. The S.208, a five-seater based on S.205 components, flew for the first time in May 1967. In addition to an order for 40 from the Italian Air Force (S.208M), the S.208 was also delivered to customers in Europe and Africa. A version designated S.208AG has been developed for general duties and features a fixed undercarriage. Maximum cruising speed is 300km/h. *Country of origin:* Italy. *Silhouette:* S.208. *Picture:* S.208M.

Power: 1 × Lycoming piston engine   *Span:* 10.67m   *Length:* 7.06m

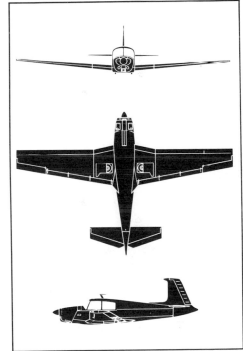

The four-seat all-metal Mooney Ranger is one of several Mooney light aircraft featuring the distinctive fin and rudder with its unswept leading edge. Originally known as the Mark 21, the Ranger first flew in 1961, although the earlier M-20 had flown in 1953. Other models in the series are the Executive, with a 200hp Lycoming, and the Mooney 201 (M20J), which first flew in 1976. The Mark 22 is a five-seat pressurised version which first flew in 1964. In mid 1986 the Mooney 205 appeared, which is developed from the 201. Another variant is the 252TSE (Turbo Special Edition). An increased length fuselage is fitted to the M20L. *Country of origin:* USA. *Silhouette:* Ranger. *Picture:* Mooney 201.

Gardan/Socata GY-80 Horizon

Confusion: Ranger, Picchio

Power: 1 × Lycoming piston engine   *Span:* 9.70m   *Length:* 6.64m

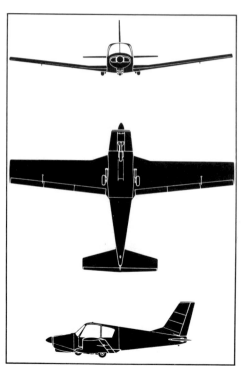

Designed by Yves Gardan, this four-seat all-metal light aircraft was put into series production by Sud Aviation from 1963. Four versions with engines of different power were originally offered. By mid-1968 more than 260 Horizons had been delivered to customers, mainly in Europe. Design and construction are conventional and the tricycle undercarriage retracts rearwards, with the wheels half concealed. Powered by a 160hp Lycoming, the Horizon has a cruising speed of about 233km/h. Other performance figures include a sea-level rate of climb of 201m/min and range with 200 litres fuel of 950km. *Country of origin:* France.

Power: 1 × Continental piston engine   *Span:* 10.41m   *Length:* 8.02m

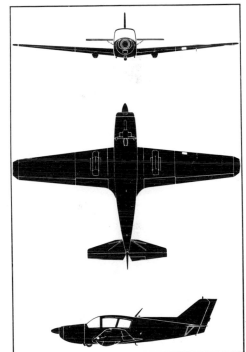

The Viking series of light aircraft consists of three models developed from the earlier Bellanca 260C and Standard Viking 300: the Model 17-30A Super Viking 300A, powered by a 300hp Continental; Model 17-31A Super Viking 300A, identical to the Model 17-30A with the exception of engine and propeller; and the Model 17-31ATC Turbo Viking 300A, which has two Rajay turbochargers added to its 300hp Lycoming. Total production of all models of Viking had reached more than 1,500 by the beginning of 1978. Production ceased in 1980. Accommodation is provided for four persons and top speed of all models is more than 357km/h. *Country of origin:* USA. *Silhouette:* Viking 300. *Picture:* Super Viking.

Beechcraft Bonanza 33/36

Confusion: Bonanza V35

Power: 1 × Continental piston engine   *Span:* 10.21m   *Length:* 8.13m (F33): 8.38m (A36)

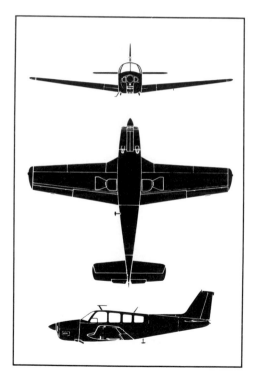

The Bonanza Model F33 is a four/five-seat executive aircraft, similar to the Bonanza Model V35 but featuring a conventional tail unit with swept fin and rudder. Originally known as the Debonair, it first flew in September 1959. An aerobatic version was built and is in service as a trainer in several air forces. Other Model F33s are used as trainers by US and European airlines. Production of the Model 33 had reached more than 2,750 by early 1989. The Bonanza Model A36 is a six-seat utility version and, like the Model 33, is powered by a 300hp Continental. Slightly longer than the Model 33, it is distinguished by an additional cabin window on each side of the rear fuselage. Production had reached more than 2,840 by early 1988. The Bonanza 36 has a maximum cruising speed of 340km/h. *Country of origin:* USA. *Silhouette:* A36. *Picture:* A36TC.

Power: 1 × Continental piston engine   *Span:* 10.21m   *Length:* 8.05m

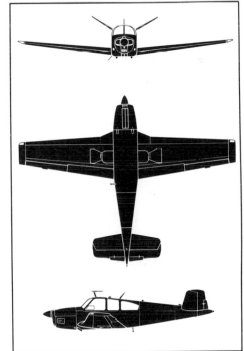

This four/five-seat V-tail light aircraft first flew in December 1945 and when production ceased in 1985, 10,000 V-tails had been built. Powered by a 285hp Continental driving a two-blade constant-speed propeller, the V35B has a cruising speed of about 322km/h. Externally the V35 Bonanza has changed little over the years and is immediately identifiable by the distinctive arrangement of the tail. Internal refinements continue to mirror changing tastes and advances in avionic equipment. *Country of origin:* USA. *Silhouette and picture:* V35B.

Avions Pierre Robin ATL Club *Confusion:* Bonanza

Power: 1 × Converted Volkswagen piston engine   *Span:* 10.25m   *Length:* 6.72m

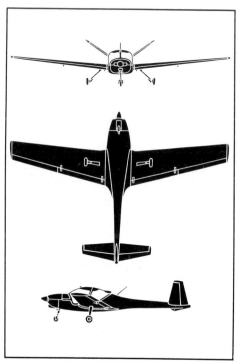

France has consistently produced good, cheap, and economical to operate light club aircraft. The two-seat Robin ATL Club follows this pattern and is in service in considerable numbers. Powered by a converted Volkswagen motor car engine, the ATL Club is unusual in having forward swept wings, a very slim rear fuselage and a V-tail. Beyond the basic version (known in the United Kingdom as Bijou) there are developments including Model 88, the ATL Voyage with up-rated engine, ATL2+2 with four seats and a re-engined version for the German market. The basic ATL Club has an economical cruising speed of 142km/h. *Country of origin:* France.

Piper PA-28R-200 Arrow

Power: 1 × Lycoming piston engine   *Span:* 9.82m   *Length:* 7.5m

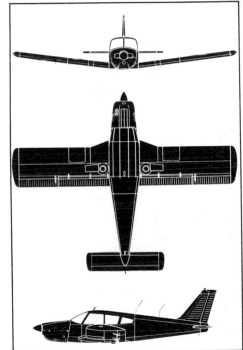

The Cherokee Arrow II is generally similar in appearance to the Cherokee Archer II but has a retractable undercarriage, more powerful engine and the untapered wings of the PA-28-180 Archer. First flown in 1975, it was superseded in 1977 by the PA-28R-201 Arrow III, which has the new longer-span tapered wings. Third in the Arrow series is the Turbo Arrow III, which is identical to the Arrow III but has a 200hp turbocharged Continental engine driving a two-blade constant-speed propeller. The Turbo Arrow III is a few inches longer than the standard aircraft and has a remodelled cowling with a larger airscoop and more pointed spinner. The Arrow IV (see page 359) has a T-tail. *Country of origin:* USA. *Silhouette and picture:* Arrow II.

Robin HR 100

Confusion: Arrow

Power: 1 × Lycoming piston engine   *Span:* 9.08m   *Length:* 7.59m

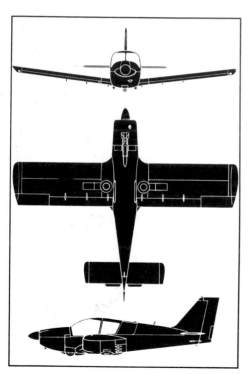

There are three basic models in the HR 100 series of French light aircraft. The fixed-undercarriage four-seat HR 100/210 Safari II is a development of the original all-metal production HR 100/200, of which 31 were built. It first flew in 1971 and remained in production until 1976, some 78 having been built. The retractable-undercarriage four/five-seat HR 100/285 Tiara first flew in 1972 and featured a new vertical tail and wing structure. Powerplant was a 285hp Lycoming driving a three-blade constant-speed propeller. The HR 100/250TR is similar to the Tiara but has a lower-powered engine. Third of the basic models is the six-seat HR 100/4+2, which first flew in May 1975. It is 0.3m longer than the Tiara and features an additional set of cabin windows. Powerplant is a 320hp Lycoming driving a three-blade propeller. Maximum cruising speed is 270km/h. *Country of origin:* France. *Silhouette:* HR 100 Tiara. *Picture:* HR 100/280.

Power: 1 × Lycoming piston engine　　*Span:* 10.85m　　*Length:* 7.63m

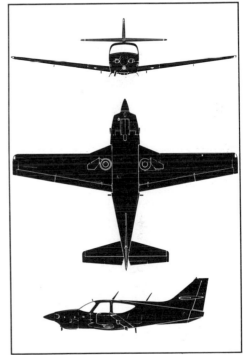

These attractive four-seat light aircraft are all basically similar, the differences lying in equipment and powerplants. The Commander 112 first flew in December 1970, with production aircraft being delivered from 1972. Successive models were improved in exterior and interior detail but retained the same exterior shape. The Commander 114 was introduced in 1976 and differs from the Model 112 principally in having a 260hp engine rather than the 112's 210hp unit. All aircraft have the characteristic mid-set tailplane and sharply tapered rear cabin windows. Cruising speed is 281km/h. Other performance figures include a maximum rate of climb at sea level of 305m/min, and a range with maximum fuel, no reserves, of 1,818km. *Country of origin:* USA. *Silhouette:* Commander 112. *Picture:* Commander 114.

Piaggio P.149 *Confusion:* Guepard

Power: 1 × Lycoming piston engine   *Span:* 11.12m   *Length:* 8.8m

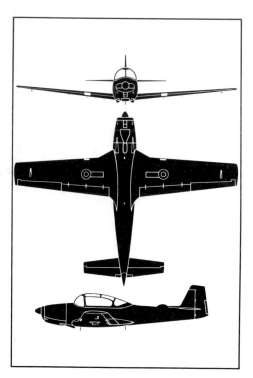

The P.149, developed from the P.148, is a four/five-seat all-metal light aircraft. First flown in 1953, the type was built in quantity as a liaison and training aircraft for the Federal German Air Force and was subsequently built under licence by Focke-Wulf in Germany. Powered by a 270hp Lycoming driving a three-blade constant-speed propeller, the P.149D cruises at 266km/h. Other performance figures include a maximum level speed at sea level of 309km/h and range (including allowance for starting, warm-up, take-off and climb, plus 30 min reserves) of 1,095km. *Country of origin:* Italy. *Silhouette and picture:* P.149D.

Wassmer WA-40/Super IV/CE.43 Guepard

Power: 1 × Lycoming piston engine   *Span:* 10m   *Length:* 8.09m

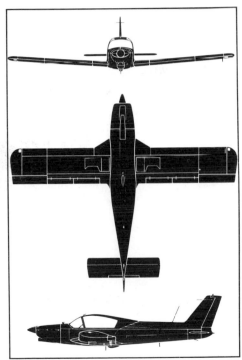

Several versions of the Wassmer WA-40 four/five-seat cabin light aircraft have been built since the first flew in 1959. Later aircraft introduced a swept fin and rudder (WA-40A) and improved engine cowling. The Super IV/21 is externally similar to the Super IV but is powered by a 250hp Lycoming driving a constant-speed propeller. The Cerva CE.43 Guepard is an all-metal derivative of the Super IV/21 and first flew in 1971. Production ended in 1976 after 43 had been built. Wassmer went into liquidation in 1977 and responsibility for Wassmer and Cerva products passed to Issoire-Aviation. *Country of origin:* France. *Silhouette:* WA-40A. *Picture:* Super IV.

SIAI-Marchetti SF.260

Confusion: Guepard, Redigo

Power: 1 × Lycoming piston engine   *Span:* 8.35m   *Length:* 7.1m

The SF.260 has been built in both civil and military versions since it first entered production in 1964. Derived from the Aviamilano F.250, it is of conventional construction and is fully aerobatic. Military versions are the SF.260M two/three-seat trainer, first flown in 1970, and the SF.260W Warrior weapons trainer/tactical support aircraft, first flown in 1972. These versions are fitted with two or four underwing stores pylons. Both types are widely deployed throughout the world. Civil versions are the SF.260A, B and C. The SF.260A was marketed in the US as the Waco Meteor. Outwardly similar to the military versions, civil SF.260s are also widely distributed throughout the world. *Country of origin:* Italy. *Silhouette:* SF.260. *Picture:* SF.260M.

Power: 1 × Allison 250-B17 turboprop   *Span:* 10.34m   *Length:* 7.9m

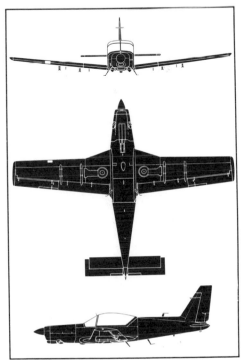

A primary/basic trainer, the two-seat Redigo first flew in July 1986. It is developed from the earlier Valmet L-70 and differs in being larger, having different wings, retractable undercarriage and a turboprop engine. The Finnish Government is purchasing 10 Redigos but these will be used as liaison aircraft rather than as trainers. The Redigo can be fitted with bombs, rockets or gun pods. Maximum level speed is 335km/h and endurance more than 5 hours. *Country of origin:* Finland.

HAL HPT-32 *Confusion:* SF.360

Power: 1 × Lycoming piston engine   *Span:* 9.5m   *Length:* 7.7m

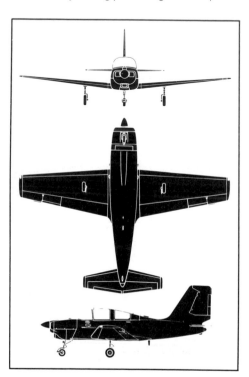

With side-by-side forward seats and an additional seat behind, the HPT-32 is the standard basic trainer for the Indian Air Force first flown in January 1977. Some 80 aircraft have been ordered to date plus 8 for the Indian Navy. Four underwing store points are provided for weapon training, and the type can also be used for observation, liaison and sports flying. Maximum speed is 217km/h and endurance 4h. *Country of origin:* India.

Power: 1 × Lycoming piston engine   *Span:* 9.98m   *Length:* 7.85m

Originally known as the Musketeer Super R, this type was first introduced in 1969. In 1971 the manufacturer renamed the entire Musketeer range and the Super R became the Sierra 200. Since then several detail improvements have been made, including the fitting of a more powerful engine and general aerodynamic cleaning-up. With accommodation for four to six persons, the Sierra can be identified by its four cabin windows on each side (against three in the similar but fixed-undercarriage Sundowner 180). Powered by a 200hp Lycoming flat-four, the Sierra has a cruising speed of about 257km/h. *Country of origin:* USA.

Bellanca Model 25 SkyRocket II

Confusion: Sierra 200

Power: 1 × Continental piston engine   *Span:* 10.67m   *Length:* 8.81m

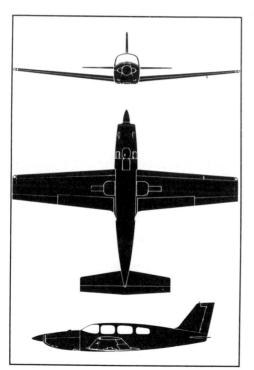

Design of this sleek six-seat executive aircraft was begun as long ago as 1956 by the late G. M. Bellanca. The choice of innovative materials for the type's construction lengthened development time, but it is claimed that the glass-fibre epoxy laminate used has a higher strength/weight ratio than aluminium and is also aerodynamically extremely smooth. The prototype first flew in March 1975 and by 1978 orders for some 70 aircraft had been recorded. Cruising speed is 410km/h. Other performance figures include a maximum cruising speed (83 per cent power) of 532km/h and a range at 4,570m and 75 per cent power of over 1,955km. The Bellanca Aircraft Corp. assets were acquired, in 1982 by Viking Aviation. *Country of origin:* USA.

Power: 1 × Ivchenko or 1 × Vedeneev piston engine *Span:* 10.6m *Length:* 8.35m

Several versions of this Russian two-seat basic trainer have been built since it first appeared in 1946. The original aircraft was a tailwheel-undercarriage type with a five-cylinder radial engine closely cowled with individual cylinder helmets. Introduced in 1957, the Yak-18A was a cleaned-up version of the Yak-18U with a tricycle undercarriage and a more powerful nine-cylinder radial engine. Widely used as a club aircraft, the Yak-18A has been exported in large numbers throughout the Communist world. A further development is the single-seat Yak-18P 'Moose'. A four-seat tourer development is the Yak-50, used extensively for Aeroflot training. Developments are the Yak-50 and the Yak-52 and -53 with tricycle undercarriage. *Country of origin:* USSR. *Silhouette and picture:* Yak-18P.

North American T-6 Texan

Confusion: Yak-18

Power: 1 × R-1340 piston engine   *Span:* 12.8m   *Length:* 8.99m

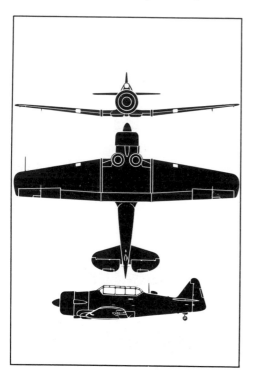

The extremely long-lived T-6 Texan was first produced before the Second World War and was widely used as an advanced trainer in that conflict, more than 10,000 being built in North America. The type was known as the Harvard in RAF service. The Texans to be seen today are virtually the same as the wartime aircraft. The type is still in wide use in South Africa, South America and elsewhere as an air force trainer and racing aircraft. *Country of origin:* USA. *Silhouette:* T-6G. *Picture:* SNJ-3 (naval version).

Power: 1 × T53 turboprop   *Span:* 12.19m   *Length:* 10.26m

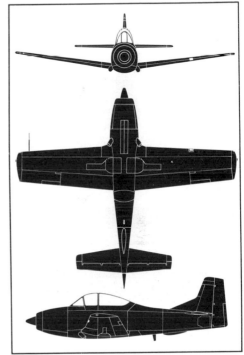

This tandem two-seat trainer was designed to fill a Nationalist Chinese Air Force requirement for an advanced trainer/ground attack aircraft. It first flew in November 1973 and entered series production in May 1976, 50 having been built before production ceased in late 1981. Powered by a licence-built 1,450hp Lycoming turboprop, the T-CH-1 cruises at more than 402km/h. Externally, the design owes much to the North American T-28A, particularly in the arrangement of the cockpit and tail assembly. Cruising speed is 315km/h. *Country of origin:* Taiwan.

North American T-28 Trojan

Confusion: T-CH-1

Power: 1 × R-1820 piston engine   *Span:* 12.38m   *Length:* 10m

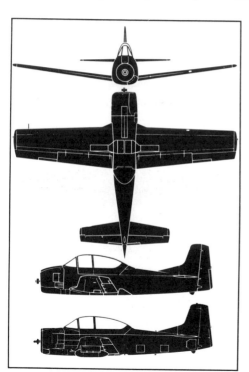

Originally designed as a replacement for the T-6 Texan, this tandem two-seat trainer first flew in September 1949. First operated by the US Air Force and US Navy, the type has since been used by a number of air forces throughout the world. The T-28A, T-28B and T-28C were all trainers, but the T-28D, slightly larger and fitted with a more powerful engine, has been widely used in the counter-insurgency role. A total of six underwing stations can accommodate a variety of armament. About 50 T-28s of all versions can still be seen in South America and South-east Asia, as well as in the US. *Country of origin:* USA. *Main silhouette:* T-28C; *lower side view:* T-28D. *Picture:* T-28D.

Piper PA-46-310P Malibu

Power: 1 × Continental or Lycoming piston engine   *Span:* 13.11m   *Length:* 8.66m

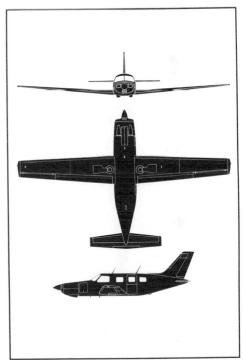

Claimed to be the world's first pressurised cabin aircraft powered by a single piston engine, the Malibu is a large six-seater able to cruise at heights up to 7,620m. First deliveries were made in November 1983 and several hundred have been produced. Powered by a 310hp turbocharged Continental engine, the Malibu cruises at up to 398km/h. Latest version is powered by a Lycoming engine and is called the Malibu Mirage. *Country of origin:* USA.

Piper Lance I/Lance II/Turbo Lance II

Confusion: (Lance): Cherokee

Power: 1 × Lycoming piston engine   *Span:* 10m   *Length:* 8.44m

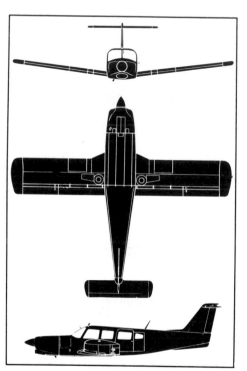

There are three main versions of the Piper Lance. The original model, first flown in August 1974, was essentially a retractable-undercarriage Cherokee Six and featured a conventional tail unit. The Lance II and Turbo Lance II both have the now fashionable T-tail arrangement, while the Turbo Lance also has the extra power of a 300hp turbocharged Lycoming engine. The T-tail is claimed to reduce drag and to increase performance. Main distinguishing point of the Turbo Lance is the large air inlet for the turbocharger located beneath the propeller spinner. *Country of origin:* USA. *Silhouette:* Turbo Lance II. *Picture:* Lance II.

Power: 1 × Lycoming piston engine   *Span:* 10.8m   *Length.* 8.23m

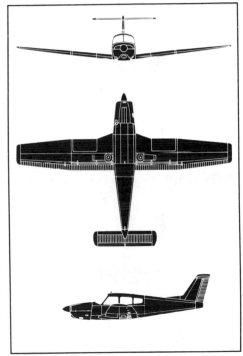

Developed from the fixed-undercarriage Archer II, the Arrow IV is a four-seat monoplane with a T-tail and retractable tricycle undercarriage. The T-tail was introduced in 1979. Production of the Lycoming engined Arrow IV ceased in 1982 and it was replaced on the line by the Turbo Arrow IV powered by a turbocharged Continental engine. Over 1,000 Turbo Arrow IVs have been built. The Arrow IV has a cruising speed of 311km/h and a range of 1,353km. *Picture:* Turbo Arrow IV. *Country of origin:* USA.

Breguet BR 1050 Alizé

Confusion: —

Power: 1 × Dart turboprop   *Span:* 15.6m   *Length:* 13.86m

The three-seat Alizé (Tradewind) is a carrier-based anti-submarine aircraft currently in service with the navies of France and India. First flight was in 1956 and two prototypes and three pre-production aircraft preceded 75 production Alizés into *Aéronavale* service from mid-1959. Armament includes a torpedo or depth charges, AS.12 air-to-surface missiles or rockets. The Alizé has a large retractable radome underneath the rear fuselage. Noticeable also are the large landing gear/weapons nacelles on the wing leading edges. *Country of origin:* France.

Ayres Turbo-Thrush

SOCATA TB 10 Tobago

Gulfstream Aerospace T-Cat

Power: 1 × Lycoming piston engine   *Span:* 9.98m   *Length:* 7.85m

The Beechcraft Musketeer series began with the Model 23 Musketeer, which first flew in 1961. The series has been progressively improved over the years, although the basic exterior appearance has remained largely unchanged. In 1974 Beech renamed the series as follows: the original Musketeer Super R became the rectractable-undercarriage Sierra 200; the Musketeer Custom became the Sundowner 180; and the Musketeer Sport became the Sport 150. These new designations reflect the engine power. Main external difference between the Sundowner 180 and the Sport 150 is the former's extra cabin window on each side. Cruising speeds are about 225km/h for the Sundowner and 200km/h for the Sport. *Country of origin:* USA. *Silhouette:* Sierra 200. *Picture:* Sundowner 180.

Gulfstream Aerospace AA-5A/AA-5B

Power: 1 × Lycoming piston engine   *Span:* 9.6m   *Length:* 6.71m

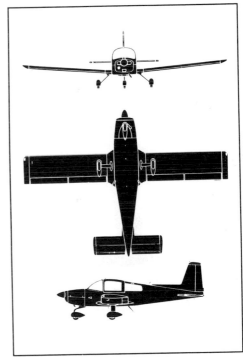

A larger development of the AA-1 series, the AA-5 first flew in 1970. This all-metal four-seat aircraft was built in four versions; the basic AA-5A with 150hp Lycoming engine; the de luxe Cheetah, externally identical to the AA-5A but with additional internal equipment; the AA-5B with 180hp Lycoming; and the de luxe Tiger, which has the same internal equipment as the Cheetah. The AA-5B series first flew in 1974. The Cheetah was known as the Traveler until 1976. Cruising speeds are 235km/h for the AA-5A and 260km/h for the AA-5B. *Country of origin:* USA. *Silhouette:* AA-5A. *Picture:* Cheetah.

Gulfstream Aerospace AA-1/T-Cat/Lynx

Confusion: AA-5

Power: 1 × Lycoming piston engine   *Span:* 7.46m   *Length:* 5.86m

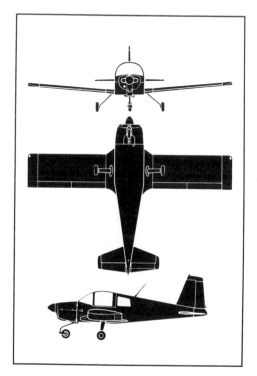

Designed originally as a trainer version of the American Aviation AA-1 Yankee, the AA-1A first flew in 1970. It was re-designated AA-1B Trainer in 1973. The AA-1C was introduced in 1977 and incorporated a number of minor design changes as well as a more powerful 115hp Lycoming engine. The T-Cat is externally identical to the AA-1C but includes a number of internal equipment changes. The Lynx is designed to be both a trainer and sports aircraft and is equipped with a de luxe interior and exterior finish, and wheel fairings. Cruising speed of the T-Cat and Lynx is about 210km/h. *Country of origin:* USA. *Silhouette and picture:* AA-1 Trainer.

Aerospace of New Zealand Airtrainer CT4

Power: 1 × Continental piston engine   *Span:* 7.92m   *Length:* 7.06m

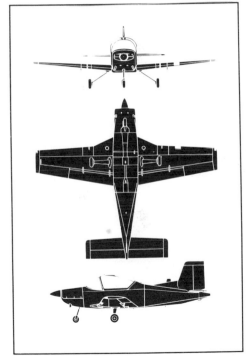

The original Victa Airtourer was a two-seat all-metal fully aerobatic light aircraft which first flew in 1959. Production aircraft were powered by a 100hp Continental engine. A four-seat development, the Aircruiser, was flown in 1967 but not produced in quantity. Aerospace of New Zealand acquired the Australian company's Airtourer and Aircruiser designs in 1971 and re-engineered the latter into a two/three-seat military trainer. Called the Airtrainer CT4, it is powered by a 210hp Continental and cruises at about 260km/h. A small number of Airtrainers and Airtourers serve with the Royal New Zealand Air Force. *Country of origin:* New Zealand.

UTVA-75A21 *Confusion:* Airtrainer

Power: 1 × Textron Lycoming piston engine   *Span:* 9.73m   *Length:* 7.11m

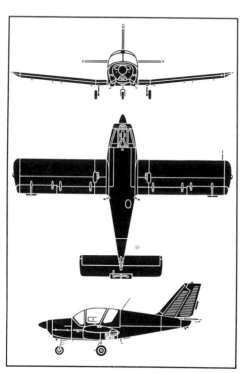

Used by both the Yugoslav Air Force and flying clubs, the UTVA-75A21 is a two seat side-by-side trainer and utility aircraft. It first flew in May 1986 and more than 100 have been delivered. Weapons can be carried on underwing racks. The aircraft has a maximum speed of 215km/h and a range of 800km. A four seat version is the -75A41. *Country of origin:* Yugoslavia.

AIDC PL-1B Chiensou

Power: 1 × Lycoming piston engine   *Span:* 8.53m   *Length:* 5.99m

The PL-1B is a Taiwan-built version of the PL-1 Laminar, an amateur-designed two-seat light aircraft. First flown in 1962, this conventional tricycle-undercarriage all-metal light aircraft was adopted for use by the Taiwanese Air Force after an example was built and flown at Taichung in 1968. The production version is designated PL-1B Chiensou; a total of 55 were built. Powered by a 150hp Lycoming piston engine, the PL-1B cruises at about 210km/h. Other variants of the basic Pazmany PL-1 include the PL-2, with wing dihedral increased from 3° to 5°, wider cockpit and extensively changed internal structure to simplify construction, and the Indonesian-built Lipnur LT-200. *Country of origin:* Taiwan/USA.

Robin HR 200

Confusion: DR 400

Power: 1 × Lycoming piston engine   *Span:* 8.33m   *Length:* 6.64m

Although the all-metal two-seat HR 200 series bears an external resemblance to the HR 100/210 it is a separate design. Slightly smaller overall, it is intended for schools and clubs. First flight was in 1971 and there are four versions: HR 200/100S, basic version; HR 200/100 Club, similar to the HR 200/100 with 108hp Lycoming engine but with wheel fairings and Hoffmann propeller; HR 200/120B with 118hp Lycoming engine; and HR 200/160 with 160hp Lycoming engine. Main distinguishing feature is the large forward-sliding clear canopy giving almost all-round vision. Cruising speed ranges from about 195km/h for the HR 200/100 to 250km/h for the HR 200/160. *Country of origin:* France. *Silhouette and picture:* HR 200/100 Club.

Power: 1 × Lycoming piston engine   *Span:* 8.72m   *Length:* 6.96m

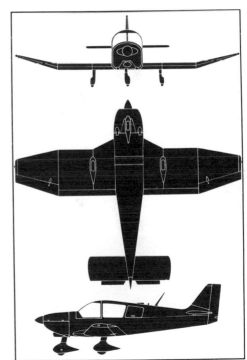

The DR 400/100 2 + 2 is a development of the earlier DR 220 2 + 2 and first flew in 1972. It is powered by a 100hp engine. The DR 400/120 Petit Prince/Dauphin also flew for the first time in 1972 and is powered by a 118hp Lycoming engine. The DR 400/140B Major is powered by a 160hp Lycoming engine and is externally the same as the 2 + 2 and Petit Prince. The DR 400/600 Chevalier replaced the earlier DR 360 Major 160 and is powered by a 160hp Lycoming engine. Most powerful aircraft in the range is the DR 400/180 Regent (together with the DR 400/180R Remorqueur glider tug), which replaced the earlier DR 235 Regent and DR 380 Prince. This four/five-seater is powered by a 180hp Lycoming. Latest version is the DR 400/100 Cadet which is a two seater. *Country of origin:* France. *Silhouette and picture:* DR 400/180.

Robin DR 221 Dauphin   *Confusion:* DR 400

Power: 1 × Continental or 1 × Lycoming piston engine   *Span:* 8.72m   *Length:* 7m

The three/four-seat DR 221 Dauphin, a development of the DR 220 2 + 2 range, first flew in 1967. All of the aircraft in this range have the characteristic cranked wing of Jodel-inspired light aircraft. The basic DR 220 and the strengthened DR 220A are both powered by a 100hp Rolls-Royce Continental, while the DR 220/108 is fitted with a 115hp Lycoming. The Dauphin is powered by a 115hp Lycoming and cruises at 205km/h. Other performance figures include a maximum rate of climb at sea level of 198m/min; service ceiling of 3,900m; and maximum range of 910km. More than 140 DR 220s and DR 221s were built before production switched to later Robin designs. *Country of origin:* France. *Silhouette:* DR 220. *Picture:* DR 221.

Power: 1 × Lycoming piston engine   *Span:* 9.76m   *Length:* 7.63m

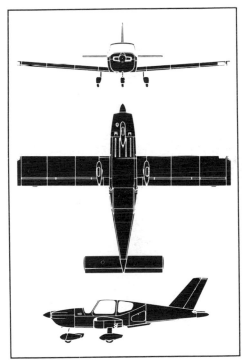

The TB 9 Tampico is a 160hp four-seater, while the TB 10 Tobago is powered by a 180hp Lycoming and is classed as a four/five seater. An aerobatic version designated TB 11 was built, but is not currently offered. The fin of the TB 9 and TB 10 is comparatively tall and is not fitted with a fillet where it joins the fuselage. A more powerful version with retractable landing gear is the TB 20 Trinidad. With turbocharger it becomes the TB 21 TC. Sales of all versions totals in excess of 500. *Country of origin:* France. *Silhouette and picture:* Tobago.

Piper PA-28/PA-32 Cherokee

Confusion: DR 400, Saratoga

Power: 1 × Lycoming piston engine *Span:* 9.14m *Length:* 7.16m

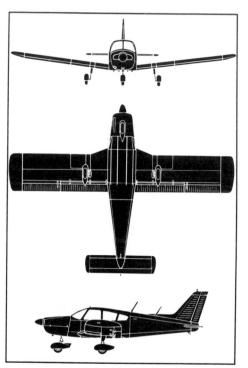

The original PA-28-140 Cherokee 140 two-seater was announced in 1964. The PA-28-180 Cherokee 180 is a four-seater powered by a 180hp Lycoming engine. Slightly longer than the 140, it has three cabin windows each side. The PA-28-235 is slightly larger than the Model 180. The introduction of the PA-28-161 Warrior in 1973 saw the first use of a new increased-span tapered wing. The 1977 versions, designated Warrior II, have the original 150hp engine replaced by a 160hp unit. The PA-28-181 Archer four-seater was introduced in 1972 and the 235hp PA-28-236 Dakota first appeared in 1978. Over 750 Dakotas have been sold. The Cherokee Six (PA-32) has a longer fuselage and four windows per side. Latest, modified, Cherokee is called Cadet. *Country of origin:* USA. *Silhouette:* Cherokee 180. *Picture:* Archer II.

Power: 1 × Lycoming piston engine   *Span:* 11.02m   *Length:* 8.44m

A large six/seven seater of typical Piper outline, the Saratoga was first introduced in late 1979. It is produced in four versions; a basic model with fixed or retractable landing gear, and two corresponding turbo-charged models. Turbo Saratogas can be identified by a prominent air intake below the propeller. Saratogas resemble the Cherokee Six in having four windows per side, but have longer-span wings. Cruising speed of the Saratoga is 267km/h. Two retractable undercarriage variants are the Saratoga SP and Turbo Saratoga SP. Over 800 of these types have been sold. *Country of origin:* USA.

Power: 1 × Lycoming piston engine   *Span:* 9.75m   *Length:* 7.5m

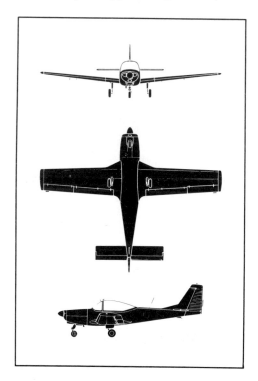

Following an agreement with the Italian company SIAI-Marchetti in 1967, FFA has developed and produced the two/three-seat AS 202 Bravo light trainer and sporting aircraft. First flown in 1969, the Bravo is currently available in two versions: AS 202/15 powered by a 150hp Lycoming engine, of which some 34 were sold and the AS 202/18A powered by a 180hp Lycoming. More than 180 Bravos had been sold by 1988. The British Aerospace Flying College uses the 18A with the name Wren. Cruising speed of the AS 202/18A is 203km/h and maximum range is 890km. *Country of origin:* Switzerland. *Silhouette and picture:* AS 202/15.

Slingsby T67M Firefly 160/T67M200 Firefly

Power: 1 × Textron Lycoming piston engine   *Span:* 10.59m   *Length:* 7.32m

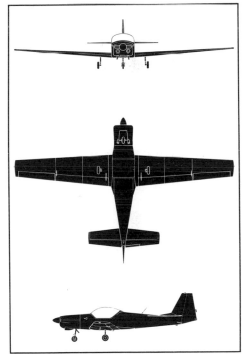

The two-seat military basic trainer version of the T67 series, which was based on the Fournier RF-6B, the T67M Firefly 160 is powered by a Textron Lycoming flat four engine. As the T67M200 it is fitted with a higher-powered engine and has a higher cruising speed. The Firefly 160 first flew in December 1982. The T67M200 has been exported to Turkey and the Netherlands. The T67M200 has a maximum cruising speed of 246km/h and a range of 926km. Civil versions are the T67B, C and D. *Country of origin:* UK.

Fournier RF-6B/RFB RS 180 Sportsman *Confusion:* Bravo

Power: 1 × Lycoming piston engine/   *Span:* 10.5m   *Length:* (RS 180): 7.15m
     1 × Continental piston engine

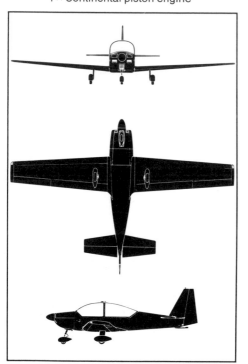

Designed by René Fournier, the four-seat RS 180 Sportsman first flew in 1973, powered by a 125hp Lycoming engine. Early production aircraft were designated RF6-180 and had the tailplane positioned on the top of the fuselage; in early 1978 the designation was changed to RS 180 and the tailplane was repositioned midway up the fin. The Fournier RF-6B, a generally similar but smaller two-seat version, first flew in 1974. Intended primarily for aerobatics and training, it has a one-piece transparent canopy and is powered by a 100hp Rolls-Royce Continental engine. Cruising speed of the RS 180 is 235km/h and that of the RF-6B about 190km/h. Production was suspended in 1981. *Country of origin:* France/West Germany. *Silhouette:* RF-6-180. *Picture:* RS-180.

Power: 1 × Textron Lycoming 0-235 piston engine   *Span:* 10m   *Length:* 7.36m

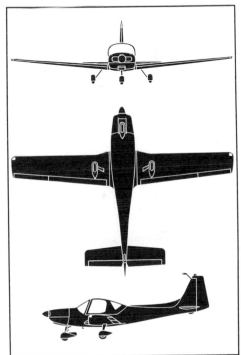

First flown in November 1985, the Grob 115 is a side-by-side two-seater which is in production at Mindelheim. The G115B has an uprated engine and a glider towing capability. By May 1988 60 115As and Bs had been ordered. The G115A has a maximum speed of 220km/h and maximum range of 1000km. Beyond this, the company has built a four seat version, the Grob G116 which first flew in April 1988. Span is increased to 11m and length to 8.50m, while maximum speed is raised to 285km/h. *Country of origin:* West Germany.

Fuji FA-200 Aero Subaru

Confusion: Vinka

Power: 1 × Lycoming piston engine   *Span:* 9.42m   *Length:* 8.17m

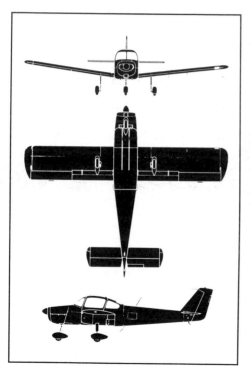

Design of this all-metal four-seat light aircraft began in 1964 and the type first flew in the following year. First production version was the FA-200-160 with a 160hp Lycoming engine. Two other versions have been built: the FA-200-180 with 180hp Lycoming engine, and the FA-200-180AO with a 180hp Lycoming and fixed-pitch propeller. Total production all versions had reached 298 by early 1982, when production ceased. *Country of origin:* Japan.

Valmet L-70 Vinka/Miltrainer

Power: 1 × Lycoming piston engine   *Span.* 9.85m   *Length:* 7.5m

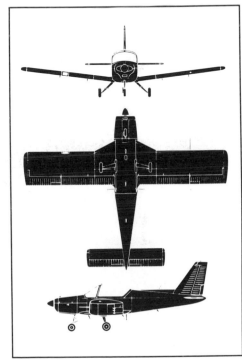

This two/four-seat training/touring light aircraft was first flown in 1975 and was produced to meet a requirement of the Finnish Air Force for a basic trainer to replace its Saab Safirs. Thirty were ordered, with deliveries completed by the end of 1982. Of conventional appearance and construction, the Vinka has a large one-piece rearward-sliding canopy and swept wing-root leading edges. Powered by a 200hp Lycoming engine, the Vinka cruises at 225km/h. The Vinka is known as the Miltrainer for export purposes. *Country of origin:* Finland.

Aerotec A-122 Uirapuru

Confusion: Vinka

Power: 1 × Lycoming piston engine   *Span:* 8.5m   *Length:* 6.6m

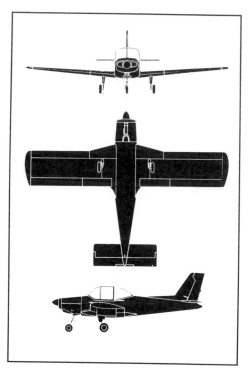

The two-seat Uirapuru first flew in 1965, powered by a 108hp Lycoming engine; it was followed by a second aircraft powered by a 150hp engine. An order for 30 A-122As was placed by the Brazilian Air Force in 1967, the aircraft being designated T-23 by the military. Powered by a 160hp Lycoming engine, the type was subsequently ordered by both Bolivia and Paraguay, while the quantity in Brazilian service was increased to 100. The civil version is designated A-122B and is similar to the A-122A but with a revised canopy. A total of 155 Uirapurus of all versions had been built by early 1977, when production ended. Cruising speed of the A-122A is 185km/h, that of the A-122B 195km/h. *Country of origin:* Brazil. *Silhouette and picture:* A-122A.

Power: 1 × Avia piston engine   *Span.* **9.11m**   *Length:* **7.07m**

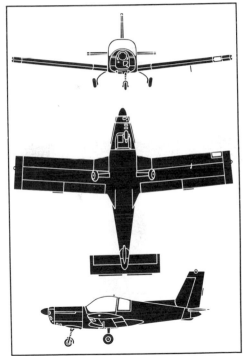

The two-seat Zlin 42, the first of a series of sporting and touring aircraft to be developed by the Moravan factory, first flew in 1967. Initial production aircraft were powered by a 180hp M137A engine and the principal operator was East Germany. Production of the Zlin 42M began in 1974. The Zlin 43, a larger, two/four-seat version of the Zlin 42, is powered by a 210hp Avia M337A. Latest version, with altered cockpit hood and more powerful engine and retractable undercarriage, is the 142. Some 250 142s have been built. *Country of origin:* Czechoslovakia. *Silhouette:* Zlin 142. *Picture:* Zlin 42M.

BAe Bulldog/Pup

Confusion: Dauphin, Zlin 42/43

Power: 1 × Continental or 1 × Lycoming piston engine

Span: (Pup): 9.45m

Length: (Pup): 6.99m

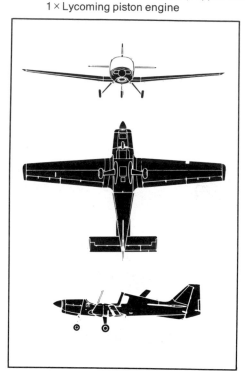

Three versions of the all-metal Pup were built. The B-121 Series 1 Pup 100, a two-seater powered by a 100hp Rolls-Royce Continental, first flew in 1967. The B.121 Series 2 Pup 150 was powered by a 150hp Lycoming engine and first flew in 1968. This version was fitted with an optional seat for a third adult or two children. Final variant was the Series 3 Pup 160, six of which were built for Iran. The Bulldog was developed in 1968 as a military trainer version and differed externally in being slightly larger and in having a fully transparent sliding canopy. It first flew in 1969 and production aircraft were fitted with a 200hp Lycoming. Customers include the RAF, Ghana, Nigeria, Jordan, Lebanon, Kenya and the Hong Kong Royal Auxiliary Air Force. Production was completed in 1982. *Country of origin:* UK. *Silhouette:* Bulldog. *Picture:* Pup 150.

Confusion: FU-24   **SOCATA Rallye**

Power: 1 × Continental piston engine   *Span:* 9.74m   *Length:* 6.97m

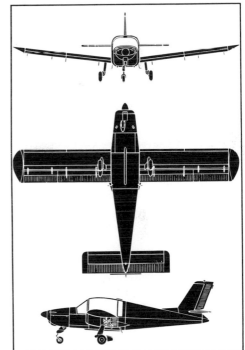

The Rallye family of light aircraft embraces several different aircraft, all originating from the Morane-Saulnier company's 90hp Rallye-Club of 1959. About 580 Rallyes were built by Morane-Saulnier. About a dozen different versions were built with engines ranging from 100hp to 235hp. Later versions were the R 235 Gabier and the armed Guerrier, operated by Rwanda and Senegal. The Rallye 100 is built under licence in Poland under the name Koliber. Cruising speeds range from 170km/h for the lowest-powered up to 245km/h for the 235hp variants. *Country of origin:* France. *Silhouette:* Rallye 100T. *Picture:* R 235 Guerrier.

PAC Fletcher FU-24/Cresco *Confusion:* Rallye

Power: 1 × Lycoming piston engine   *Span:* 12.81m   *Length:* 9.70m

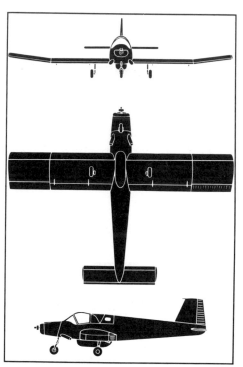

The FU-24 was initially designed by the Sargent-Fletcher company in the USA for agricultural top-dressing work in New Zealand and first flew in 1954. All manufacturing and sales rights were acquired by Air Parts in New Zealand in 1964 after 70 aircraft had been built in the US. Subsequently the type entered quantity production at Hamilton Airport, New Zealand. Customer countries include Australia, Bangladesh, Iraq, Pakistan, Thailand and Uruguay. Powered by a 400hp Lycoming, the FU-24 has an operating speed of 170–210km/h. A new version of the FU-24, the Cresco 08-600, is powered by an Avco Lycoming LTP101 turboprop and has many components in common with the standard aircraft. *Country of origin:* New Zealand. *Silhouette:* FU-24. *Picture:* Cresco.

Power. 1 × Lycoming piston engine   *Span:* 9.14m   *Length:* 7.32m

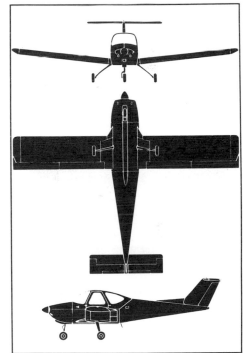

The prototype of this two-seat light trainer, designated PD 285, first flew in 1975, fitted at that time with a conventional tail. Production aircraft, designated Model 77, have the now fashionable T-tail. Some 304 Skippers had been delivered, mainly to Beech Aero Centres, by 1981, when production was suspended. The Skipper bears a marked resemblance to the Piper Tomahawk. *Country of origin:* USA.

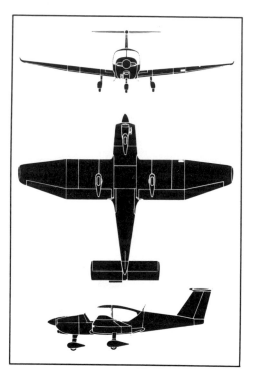

Robin R 3000

Confusion: Skipper, Tomahawk

Power: 1 × Lycoming piston engine   *Span:* 9.81m   *Length:* 7.51m

One of the later in the Robin series of light aircraft, the R 3000 is available with 116hp and 140hp engines. The second prototype, flown in June 1981, introduced the distinctive upturned wingtips. Unlike the Skipper or the Tomahawk, which also have high-set tailplanes, the R 3000 is fitted with streamlined undercarriage fairings. When fitted with the 116hp Lycoming 0-235 engine, maximum cruising speed is 215km/h. *Country of origin:* France.

Power: 1 × Lycoming piston engine   *Span:* 10.36m   *Length:* 7.06m

First introduced in 1978, this two-seat trainer followed the fashion of mounting the tailplane high on the fin. The cabin is also arranged for maximum all-round visibility by cutting down the rear fuselage decking. The Tomahawk bears a marked resemblance to Beechcraft's Skipper, but the Piper aircraft has a more substantial-looking vertical tail and lacks the small dorsal fin fairing of the Beechcraft. In 1981 the Tomahawk II was introduced, which incorporated internal changes. Production reached 2,497 of both versions before manufacture was suspended. *Country of origin:* USA.

DHC DHC-1 Chipmunk   *Confusion:* HT-2

Power: 1 × de Havilland piston engine   *Span:* 10.47m   *Length:* 7.75m

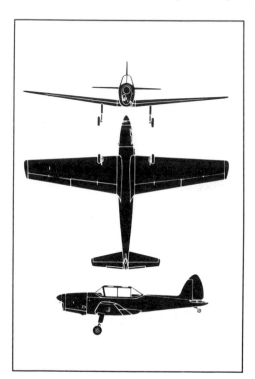

Designed by the Canadian de Havilland company, the tandem two-seat Chipmunk first flew in 1946. Some 220 were produced in Canada for both civil and military users; a further 60 were produced under licence by OGMA in Portugal. Production in the UK consisted of T.10s for the RAF and Mk 21s and 23s for civil users; Mk 22s are refurbished ex-RAF aircraft for civil use. The Mk 23 was a single-seat agricultural aircraft and the Mk 22A was a Bristol tourer version with a one-piece blown hood and wheel spats. Powered by a 145hp de Havilland Gipsy Major 10, the Chipmunk has a maximum cruising speed of 230km/h. Range is 730km. *Country of origin:* Canada. *Silhouette and picture:* T.10.

Power: 1 × Blackburn piston engine   *Span:* 10.72m   *Length:* 7.53m

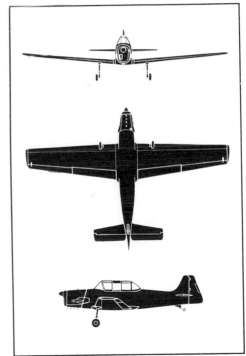

The tandem two-seat HT-2 was the first Indian powered aircraft of indigenous design and first flew in 1951. About 160 have been built, mostly for the Indian Air Force, Navy and civil users. In addition to two aircraft presented to Indonesia and Singapore, twelve were supplied to Ghana. Similar in appearance to the DHC-1 Chipmunk, though with a more angular fin and rudder, the HT-2 is powered by a 155hp Blackburn Cirrus Major III and cruises at about 185km/h. *Country of origin:* India.

Mudry CAP 10/CAP 20/CAP 21

Confusion: Emeraude

Power: 1 × Lycoming piston engine   *Span:* (CAP 10): 8.06m   *Length:* (CAP 10): 7.16m

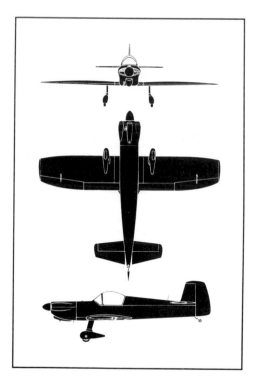

Developed from the Piel Emeraude, the two-seat all-wood CAP 10 is intended for use as a training, touring or aerobatic aircraft. First flight was in 1968, and a total of 230 CAP 10/10B have been built, including 30 for the French Air Force. The CAP 20 is essentially a single-seat derivative of the CAP 10, although almost completely new design. It has been refined considerably since it first flew in 1969, and the latest version, the CAP 20L, is suitable for competition aerobatics. Production was terminated in 1981 to make way for the CAP 21, with a new wing and undercarriage. Powered by a 200hp Lycoming engine, the CAP-20L cruises at 265km/h, compared with 250km/h for the 180hp CAP 10B. *Country of origin:* France. *Main silhouette:* CAP 20. *Picture:* CAP 20L.

Piel/Scintex CP 301 Emeraude

Power: 1 × Continental piston engine   *Span:* 8.04m   *Length:* 6.3m

Designed by Claude Piel, the two-seat, all-wood Emeraude was first flown in 1952, with the first production aircraft flying the following year. Several versions have been built, commercially and by amateurs, powered by a variety of engines from 65 to 125hp. The Scintex Emeraude is largely similar to the Piel-built aircraft but features a one-piece blown canopy in place of the upwards-opening canopy doors of the latter. The CP 301 was built under licence in small quantities in Britain, where it was known as the Fairtravel Linnet, and in West Germany by Binder Aviatik. The Super Emeraude is a slightly larger version with higher payload and improved performance. Powered by a 90hp Continental piston engine, the CP 301 Emeraude cruises at 200km/h. *Country of origin:* France. *Main silhouette:* CP 301; *lower side view:* Super Emeraude. *Picture:* Super Emeraude.

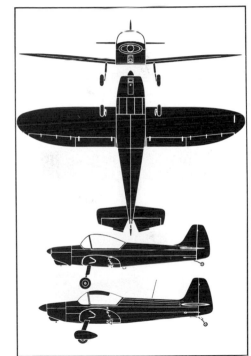

Jodel D.11/D.112/D.117

Confusion: DR 1050/1051

Power: 1 × Continental piston engine   *Span:* 8.22m   *Length:* 6.5m

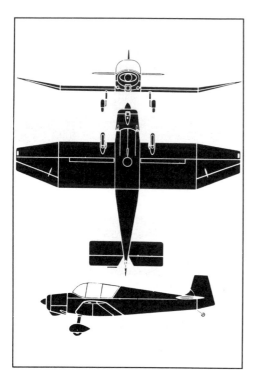

The D.11, the first two-seat light aircraft in the Jodel series, was developed from the D.9 Bébé. It first flew in 1950. The type has been built extensively by amateurs throughout Europe and North America. The D.112 is more widespread and is a refined version of the D.11. It has been built by Wassmer, SAN and Alpavia, aircraft built by the last two named companies being known as D.117s. Amateur-built examples powered by the 95hp Continental C90 are usually designated D.119. There are about 1,500 commercially built aircraft in the series. Cruising speed is 195km/h. *Country of origin:* France. *Silhouette and picture:* D.11.

Jodel DR.1050/DR.1051 Ambassadeur/Sicile

Power: 1 × Continental or 1 × Potez piston engine   *Span:* 8.72m   *Length:* 6.35m

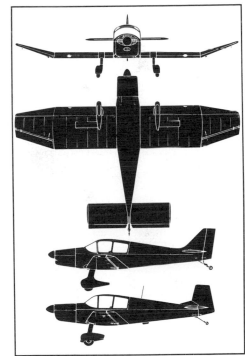

Centre Est was formed in 1957 to design and build the Jodel range of light aircraft. First in the series was the DR 100 Ambassadeur, which first flew in 1958. The first DR.1050 with a 100hp Continental was flown in 1959. A further version with a 105hp Potez engine was built in 1960 and designated DR.1051. These aircraft were also built under licence by Société Aéronautique Normande (SAN). In 1963 an improved Ambassadeur was announced, named the Sicile, with either a Continental (DR.1050) or Potez (DR.1051) engine. Last of the series were the DR.1050MM1 and DR.1051MM1 Sicile Record, with swept-back fin and rudder and one-piece elevator. Total production of the DR.100 and its derivatives amounted to more than 800 aircraft. *Country of origin:* France. *Main silhouette:* DR.1050; *upper side view:* DR.1050MM1. *Picture:* DR.1050.

Power: 1 × Continental or 1 × Potez piston engine   *Span:* 10.27m   *Length:* 7.92m

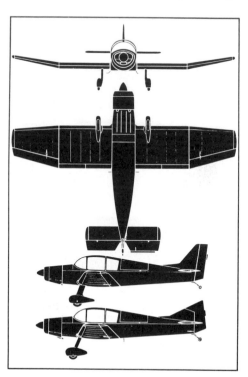

The Mousquetaire was developed by SAN as an economical four-seat light touring aircraft. It first flew in 1958 and sported a distinctive triangular fin, replaced in later aircraft with a conventional swept assembly. Five versions of the D.140 Mousquetaire were built, the sixth, the D.140R, being designated the Abeille. First flown in 1965, this aircraft was exclusively developed for glider towing and featured an extensively glazed cabin. The D.150 Mascaret is a two-seat light aircraft with a 100hp Continental or 105hp Potez. It is outwardly similar to the D.140 series but slightly smaller. Performance figures for the Mousquetaire include a maximum cruising speed at 2,300m of 240km/h; maximum sea-level rate of climb of 230m/min; and maximum-fuel range of 1,400km. *Country of origin:* France. *Main silhouette:* D.140; *lower side view:* D.140B. *Picture:* D.140.

Power: 1 × Lycoming piston engine   *Span.* 8.58m   *Length:* 6.62m

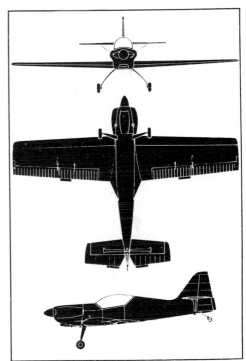

The Zlin Z 50L single-seat fully aerobatic competition aircraft first flew in 1975. The broad wings are of almost constant chord and are symmetrical in section, while the tailplane is braced on the underside. Current production versions are the Z 50 LS and LA, fitted with a more powerful 300hp Lycoming. Over 50 of this type have been built. *Country of origin:* Czechoslovakia.

ICA-Brasov IAR-822/IAR-826/IAR-827

Confusion: Pawnee, Ipanema

Power: 1 × Lycoming piston engine   *Span:* 12.8m   *Length:* 9.4m

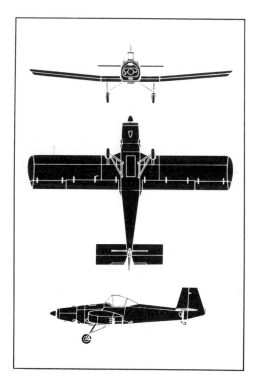

Prototype of the IAR-822 single/two-seat agricultural aircraft first flew in 1970, with series production of the first batch of 200 aircraft commencing the following year. Early aircraft were of mixed construction, but in 1973 an all-metal version was introduced, designated IAR-826. Powered by a 290hp Lycoming engine, the IAR-826 has an operating speed of 120–160km/h. The IAR-827A is a developed version of the all-metal IAR-826 with increased payload, a Polish radial engine and improved flying characteristics. Performance figures for the IAR-827 include a cruising speed of 195km/h, and maximum rate of climb at sea level of 270m/min. Latest version is known as the -828 TP. *Country of origin:* Romania. *Silhouette:* IAR-826. *Picture:* IAR-827 TP.

Power: 1 × Lycoming piston engine   *Span:* (Pawnee): 11.02   *Length:* (Pawnee): 7.53m

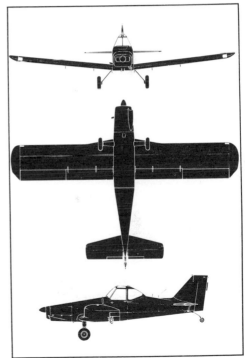

The original Pawnee single-seat agricultural aircraft first flew in 1959. The type had become one of the most used of all agricultural aircraft until production was suspended in 1982. Last production aircraft were the PA-25 Pawnee D, powered by a 235hp Lycoming engine, and the PA-36 Brave, previously known as the Pawnee Brave. Two versions of the Brave were built: the basic Brave 300 powered by a 300hp Lycoming engine and the Brave 375 with 375hp Lycoming. The Brave is externally identifiable by its swept tail assembly, lack of wing bracing and simple sprung-steel undercarriage. *Country of origin:* USA. *Silhouette:* Brave. *Picture:* Pawnee D.

EMBRAER EMB-201 Ipanema

Confusion: Pawnee, IAR-822

Power: 1 × Lycoming piston engine   *Span:* 11.69m   *Length:* 7.43m

The original version of this single-seat agricultural aircraft was designed to a specification laid down by the Brazilian Ministry of Agriculture and first flew in 1970. The following versions have been built: EMB-200 and EMB-200A, of which 73 were built between 1970 and 1974; EMB-201, the developed version with new wing leading edge and wingtips, of which 188 were built; the current-production EMB-201A, first flown in 1977 with several minor aerodynamic refinements. More than 600 Ipanemas of all versions have been sold. A glider-tug version has also been built, designated EMB-201R. Three examples have been delivered to the Brazilian Air Force. *Country of origin:* Brazil. *Silhouette:* EMB-200. *Picture:* EMB-201.

Confusion: Basant, Quail, UTVA-65

Cessna Ag Wagon/Ag Truck

Power: 1 × Continental piston engine   *Span:* 12.41m   *Length:* 8m

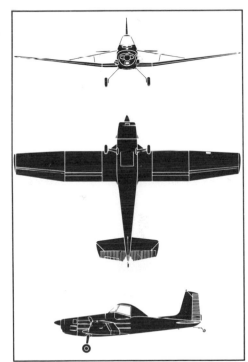

The original Model 188 Ag Wagon first flew in 1965, powered by a 230hp Continental, and incorporated a number of components common to the Model 180 high-wing aircraft of that vintage. In late 1971 Cessna introduced a new range of agricultural types, three of which were based on the earlier aircraft. One of them, the Ag Pickup, was discontinued in 1976. The Ag Wagon is powered by a 300hp Continental engine and is a conventional agricultural aircraft with strut-braced wings and enclosed cockpit with all-round vision. The Ag Truck is similar to the Ag Wagon, but has a slightly longer wing, a larger hopper and other equipment refinements. Ag Wagon production has ceased. A turbocharged version of the Ag Truck is called the Ag Husky. *Country of origin:* USA. *Silhouette:* Ag Wagon. *Picture:* Ag Husky.

437

HAL HA-31 Basant

Confusion: Ag Wagon, Privrednik

Power: 1 × Lycoming piston engine   *Span:* 12m   *Length:* 9m

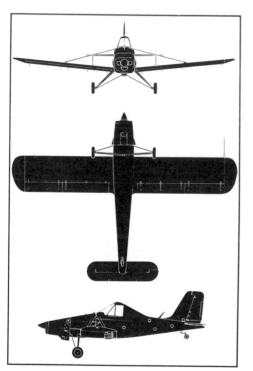

The HAL HA-31 Basant (Spring) was originally powered by a 250hp Rolls-Royce Continental but was subsequently completely redesigned as the HA-31 Mk II and re-engined with a 400hp Lycoming; it first flew in this form in 1972. A pre-production batch of 20 was built, the first eight of which were handed over to the Indian Ministry of Food and Agriculture in 1974. Series production finished in 1980 after 40 aircraft had been built. The Basant is of conventional design and construction for an agricultural aircraft, with strut-braced wings and a swept fin and rudder. Cruising speed is about 185km/h. *Country of origin:* India.

Power: 1 × Lycoming piston engine   *Span:* 10.59m   *Length:* 7.32m

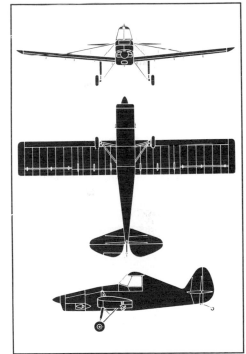

In the 1970s the Mexican AAMSA company took over production of the Aero Commander Quail and Sparrow Commander from Rockwell in the USA. The Sparrow Commander is no longer on the line but the Quail is believed to be in production under the designation A9B-M. The Quail has a normal operating speed of 145–161km/h and has a 795 litre fertiliser/insecticide hopper. *Country of origin:* Mexico. *Silhouette:* Quail Commander. *Picture:* A98-M Quail.

UTVA UTVA-65 Privrednik

Confusion: Basant, Quail, Ag Wagon

Power: 1 × Lycoming piston engine   *Span:* 12.22m   *Length:* 8.46m

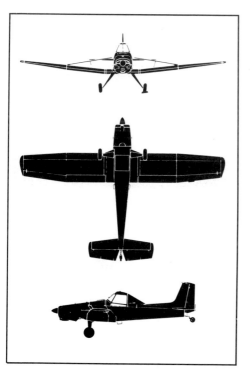

The single-seat UTVA-65 agricultural aircraft first flew in 1965 and combines the basic wings, tail assembly and undercarriage of the four-seat high-wing UTVA-60 with a new all-metal fuselage. In common with most low-winged agricultural aircraft, the UTVA-65 has a high-set, well-protected cockpit located approximately halfway along the fuselage, with a hopper forward of the cockpit. Powered by a 295hp Lycoming piston engine, the UTVA-65 has a maximum speed of 200km/h. *Country of origin:* Yugoslavia.

WSK-PZL Mielec M-18 Dromader

Power: 1 × PZL piston engine   *Span:* 17.7m   *Length:* 9.47m

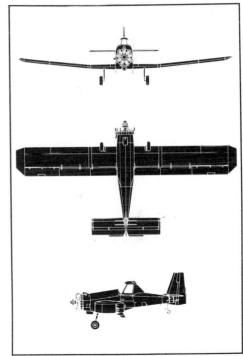

Significantly larger than the Kruk, which is also made in Poland, the M-18 Dromader first flew in 1976 and 230 had been built by early 1984. A two-seat version, designated M-18A, is also available, and a reduced capacity version known as the M-21 Mini, with a lower powered engine, has been developed. The Dromader has an operating speed of 170–190km/h and a maximum range of 520km. *Country of origin:* Poland.

PZL-Warszawa PZL-106 Kruk

Confusion: Tauro

Power: 1 × PZL piston engine   *Span:* 14.8m   *Length:* 9.1m

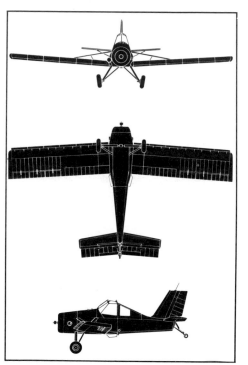

Intended as a replacement for the PZL-101 Gawron, the PZL-101M Kruk (Raven) was designed in the early 1960s. Initially powered by a 260hp Ivchenko engine, the aircraft was progressively refined into the PZL-106A. The PZL-106 first flew in 1973, powered by a 400hp Lycoming engine; production aircraft are fitted with a 600hp PZL-35 seven-cylinder radial. Production commenced in 1976. With a 1,000hp engine the Kruk is designated -106AS. The Kruk can carry its maximum chemical payload of 1,000kg at an operating speed of 120–160km/h. A further development is the PZL-106BT Turbo-Kruk, fitted with a Walter M601D turboprop engine. The Turbo-Kruk made its first flight in September 1985. *Country of origin:* Poland. *Silhouette:* PZL-106A. *Picture:* PZL-106.

Ayres Thrush/Turbo-Thrush

Power: 1 × R-1340 or 1 × R-1300   *Span:* 13.51m   *Length:* 8.95m
piston engine or 1 × PT6A turboprop

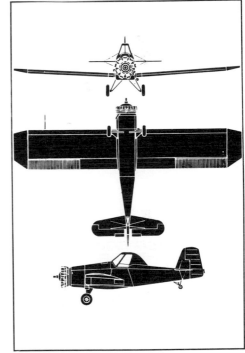

Ayres Corporation acquired the manufacturing rights to the Rockwell Thrush
Commander 600 in 1977 and has the following versions in production:
Thrush-600 with 600hp Pratt & Whitney Wasp radial and the Turbo-Thrush
S-2R, similar to the Model 600 but powered by a 750shp Pratt & Whitney
turboprop. The Pezetel Thrush has a 600hp Polish radial, and the Bull Thrush
is powered by a 1,200hp Wright Cyclone. Working speed of the Model 600 is
170–185km/h, that of the Turbo-Thrush 153–241km/h. Both the Thrush and the
Turbo-Thrush have as standard a 1.5m³ hopper capable of holding up to 1,514
litres of liquid or 1,487kg of dry chemicals. *Country of origin:* USA. *Silhouette
and picture:* Thrush-600.

Zlin Z-37 Cmelak

Confusion: —

Power: 1 × M462 piston engine   *Span:* 12.22m   *Length:* 8.55m

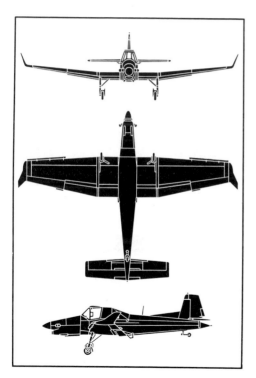

First flown in 1963, this agricultural and light utility aircraft entered series production in 1965. An improved version, the Z-37A, was introduced in 1971. It features some structural reinforcement and incorporates corrosion-resistant materials. A total of more than 700 Cmelaks (Bumble Bees) were delivered, and operators include Bulgaria, Czechoslovakia, Finland, East Germany, Hungary, India, Iraq, Poland, UK and Yugoslavia. Let has also built some 27 examples of a two-seat training version, the Z-37A-2. Latest version is the Z-37T powered by a Motorlet turboprop. This has a fin fillet and winglets at the tips. A two seat Z-37T is also available. *Country of origin:* Czechoslovakia. *Silhouette:* Z-37T. *Picture:* Z-37.

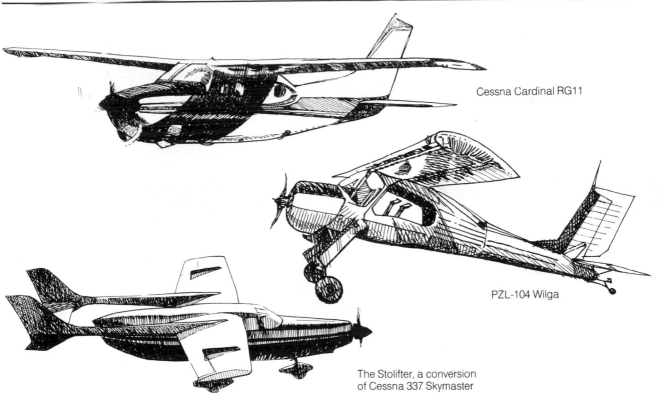

Cessna Cardinal RG11

PZL-104 Wilga

The Stolifter, a conversion
of Cessna 337 Skymaster

Piper PA-18 Super Cub

Confusion: Autocrat

Power: 1 × Lycoming piston engine   *Span:* 10.73m   *Length:* 6.88m

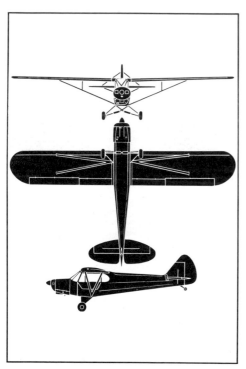

The original Super Cub, powered by a 90hp Continental, was first certificated in 1949 and was itself the successor to a long line of Piper light aircraft which began with the Taylor E-2 Cub of 1931. Later aircraft include the J-3 Cub of 1938, the PA-12 Super Cruiser of 1940 and the Vagabond, which was the forerunner of the Tri-Pacer line. Retaining the tandem two-seat arrangement, the current Super Cub is powered by a 150hp Lycoming, giving it a cruising speed of 169km/h. Most Super Cubs are used for agricultural purposes. More than 40,000 Cubs of all types have been delivered and production of the Super Cub has been resumed in complete and kit form. *Country of origin:* USA. *Silhouette and picture:* Super Cub.

Power: 1 × de Havilland or 1 × Blackburn piston engine   *Span:* 10.97m   *Length:* 7.21m

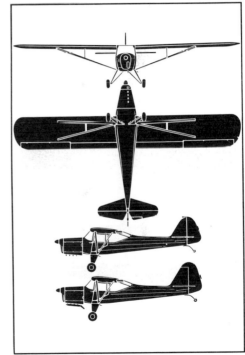

The Auster range embraces several types which can still be seen flying throughout the world. The J/1 Autocrat was the first post-war product of the Auster company and was derived from the Taylorcraft Model D spotter aircraft. Most of the range are externally similar and can be recognised by engine differences and by slight variations in the size of the vertical tail. All have V-strut wing bracing and parallel-chord wings. The Beagle A.61 Terrier is a civil conversion of the Auster AOP.6. Early conversions were also known as the Tugmaster. Introduced in 1961, the Terrier has a larger vertical tail and long exhaust pipe and silencer. *Country of origin:* UK. *Main silhouette:* AOP.6; *lower side view:* Terrier. *Picture:* Autocrat.

Aero Boero 95/115/180 *Confusion:* Citabria

Power: 1 × Continental or 1 × Lycoming piston engine   *Span:* 10.42m   *Length:* 6.90m

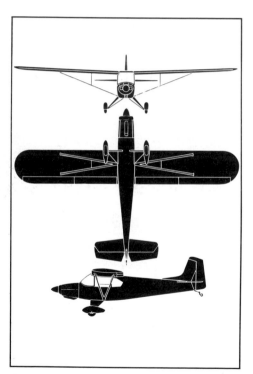

This family of three and four-seat light aircraft consists of some eleven different types: the basic AB 95 Standard with 95hp Continental; the AB 95A De Lujo with 100hp Continental; AB 95A Fumigador agricultural version; AB 95B with 150hp engine; AB 95/115 with 115hp engine, wheel fairings and streamlined engine cowling; AB 115 BS with swept-back fin and rudder; AB 180 four-seat version with 180hp Lycoming; AB 180 RV three-seat version with swept-back fin and rudder; AB 180 Condor high-altitude version with modified wingtips. Two lower-powered versions of the AB 180, the AB 150 RV and the RB 150 Ag, have also been built. *Country of origin:* **Argentina**. *Silhouette:* AB 95/115. *Picture:* AB 180.

Power: 1 × Lycoming piston engine   *Span:* (Citabria): 10.19m   *Length:* (Citabria): 6.91m

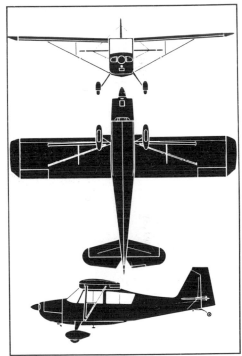

This series of high-wing light aircraft embraces several models of broadly similar appearance but widely differing powerplants and performances. Derived from the Aeronca Champion and Sedan light aircraft of the late 1940s, via the Champion Traveller and Challenger of the 1950s, current aircraft have a squared-off vertical tail and angular fuselage but retain the tandem seating of earlier aircraft. Three versions of the Citabria have been built. Production ceased in 1980. *Country of origin:* USA. *Silhouette:* Citabria. *Picture:* Decathlon.

Maule M-4 Rocket/M-5 Lunar Rocket

Confusion: Citabria

Power: 1 × Continental or 1 × Lycoming piston engine   *Span:* 9.4m   *Length:* 6.71m

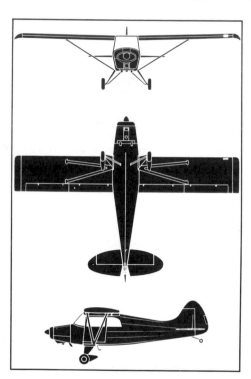

Production of the basic Maule M-4 began in 1963. This design was a conventional four-seat light aircraft with V-strut wing bracing and a rounded vertical tail. Four versions were available: the M-4 Astro-Rocket with 180hp Franklin engine; the M-4 Rocket with 210hp Continental, which was available also as a seaplane; and the M-4 Strato Rocket, generally similar to the Rocket but powered by a 220hp Franklin engine. The M-5 Lunar Rocket has 30% more flap area and enlarged tail surfaces. The five-seat M-6 Super Rocket was introduced in 1984. Powered by a 210hp Avco Lycoming, the M-5 cruises at 304km/h. Latest in the series are the MX-7 Star Rocket and the M-7-235 Super Rocket. Over 1,200 of all variants have been built. *Country of origin:* USA. *Silhouette:* M-4. *Picture:* M-5.

Power: 1 × Continental piston engine *Span:* 11.43m *Length:* 8.41m

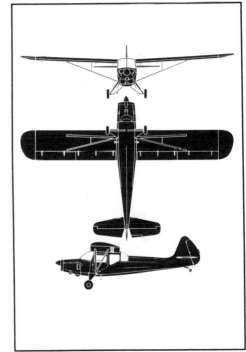

The Krishak was designed to meet an operational requirement of the Indian Air Force for an air observation post (AOP) light aircraft. It first flew in late 1959 and a total of 68 were ordered, all of which had been delivered by 1969. The Krishak normally carries a crew of two, but there is a swivelling seat for a third person in the rear of the cabin. Evolved from the HUL-26 Pushpak, the Krishak has a similar wing, with the addition of flaps, and has a more pointed vertical tail. Powered by a 225hp Rolls-Royce Continental, it has a cruising speed of 209km/h. *Country of origin:* India.

Cessna Model 180/Model 185 Skywagon

Confusion: Krishak, UTVA-60

Power: 1 × Continental piston engine *Span:* 10.92m *Length:* 7.85m

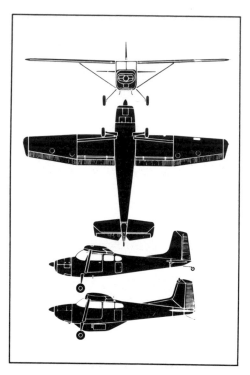

The Cessna Model 180 first flew in 1952 as a four-seater. Throughout its production life the type has retained the tailwheel undercarriage and the upright vertical tail. The Cessna 185 first flew in July 1960 and there were two versions: 185 Skywagon and the Model 185 Skywagon II, which included a factory-installed avionics package. The 185 is fitted with a 300hp Continental in place of the Model 180's 230hp unit. A military version, designated U-17 in the US, was also built. A total of 4,356 Model 185 Skywagons and U-17s had been built when production ceased in 1986. *Country of origin:* USA. *Main silhouette:* Model 180; *lower side view:* U-17. *Picture:* Model 180.

Power: 1 × Lycoming piston engine   *Span:* 11.4m   *Length:* 8.38m

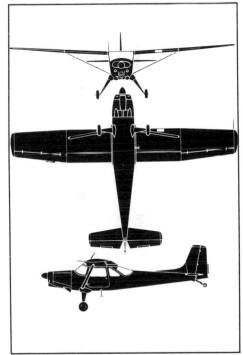

Derived from the UTVA-56 four-seat utility aircraft of 1959, the UTVA-60 is essentially the production version of the earlier aircraft. Five basic versions have been built, differing in the type of equipment carried for the various roles, which include light transport, air ambulance, agricultural spraying and float-plane. The UTVA-66 was developed from the UTVA-60 and first flew in 1967. Powered by a 270hp Lycoming engine, the UTVA-66 cruises at 230km/h. Both types are in service with the Yugoslav Air Force. *Country of origin:* Yugoslavia. *Silhouette:* UTVA-60. *Picture:* UTVA-66.

 **Cessna O-1 Bird Dog** *Confusion:* SM.1019

Power: 1 × Continental piston engine *Span:* 10.97m *Length:* 7.89m

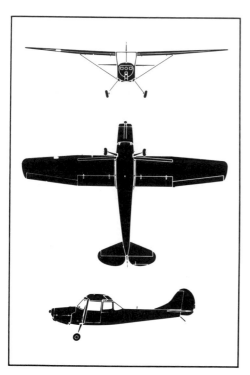

The Bird Dog was developed for the US Army as a light reconnaissance and observation aircraft which could also be used for liaison and training duties. It first flew in 1950 and remained in continuous production until late 1958, by which time more than 3,300 had been built. It briefly re-entered production in 1962 and a further 100 were built. The type is still in large-scale service in many countries throughout the world. The Bird Dog has a tailwheel undercarriage, and nearly all variants have the characteristic rounded fin and rudder. The exception is the O-1C, some 25 of which were built for the US Marine Corps, which has the old-style squared-off tail of the early Cessna 180s and 185s. *Country of origin:* USA.

Confusion: Bird Dog, AM-3C **SIAI-Marchetti SM.1019E**

Power: 1 × Allison turboprop *Span:* 10.97m *Length:* 8.52m

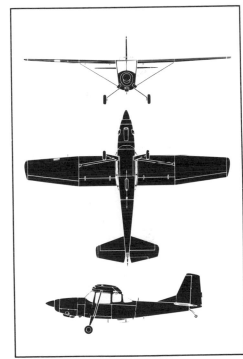

This tandem two-seat Stol light aircraft is based on the Cessna O-1 Bird Dog but has an extensively modified airframe to meet later operational requirements. First flown in 1969, it was put into production for the Italian Army, powered by a 400hp Allison 250-B17 turboprop. Total order was for 100 aircraft. The SM.1019E is stressed for two underwing hardpoints and can carry a variety of operational equipment. Cruising speed is 300km/h. Other performance figures include a maximum rate of climb at sea level of 551m/min; operational ceiling of 7,620m; and a typical operational radius with two rocket launchers of 111km. *Country of origin:* Italy.

Aeritalia AM-3C *Confusion:* SM.1019

Power: 1 × Piaggio/Lycoming piston engine   *Span:* 12.64m   *Length:* 8.93m

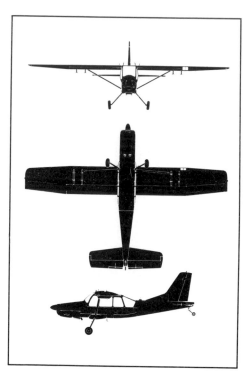

The three/four-seat AM.3C was originally designated MB.335 and first flew in 1967. Suitable for forward air control, observation, liaison, passenger and freight transport, casualty evacuation and tactical support, the type has been built in quantity for the Rwanda Air Force and the South African Air Force, in which service it is known as the Bosbok. The AM-3C is stressed to carry a variety of armaments on two underwing pylons. Powered by a Piaggio-built 340hp Lycoming, the AM-3C cruises at 245km/h. *Country of origin:* Italy.

DHC DHC-2 Beaver

Power: 1 × R-985 piston engine or/1 × PT6A turboprop   *Span:* 14.64m   *Length:* 9.24m

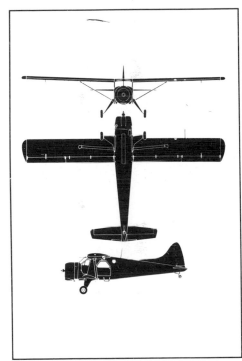

Nearly 1,700 DHC-2 Beaver light utility transports were built, the first flying in 1947. Carrying either seven passengers or freight, the Beaver has proved capable of operating in very rough, remote areas. The Beaver Mk I was produced in both civil and military forms, some 968 of the latter variant going to the US Army/USAF as the U-6A. Beavers have been fitted with floats and skis. A later development is the Turbo-Beaver Mk III, which has the Pratt & Whitney radial engine replaced by a PT6A turboprop. This changes the nose shape significantly. The Mk III can carry 10 passengers, cruises at 252km/h and has a range of 1,090km. *Country of origin:* Canada. *Silhouette:* Mk I. *Picture:* Mk III.

DHC DHC-3 Otter

Confusion: Beaver

Power: 1 × R-1340 piston engine   *Span:* 17.69m   *Length:* 12.80m

First flown in 1951, the 10-seat Otter has been extensively used by the armed forces of USA, Canada, Australia, Ghana and India, and in smaller numbers by several South American nations. The Otter can be fitted with floats and skis, and has been used in Antarctica. Powered by a 600hp Pratt & Whitney radial engine, the Otter has a cruising speed of 209km/h and a maximum level speed of 257km/h. A turboprop conversion of the Otter, powered by a single Pratt & Whitney Aircraft of Canada PT6A-27, has been produced by Cox Air Resources of Edmonton, Canada. Airtech Canada have been converting Otters to take higher rated Polish-made radial engines. Some 50 piston-engined Otters are still in service. *Country of origin:* Canada.

Max Holste Broussard

Power: 1 × R-985 piston engine   *Span:* 13.75m   *Length:* 8.65m

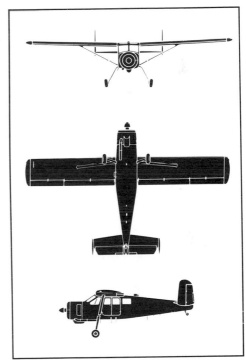

The twin-finned Broussard light utility aircraft was developed from the MH 152 experimental observation type and first flew in 1952. Production aircraft were designated MH 1521 and MH 1521A. Total production of the Broussard exceeded 330 aircraft. Over 30 remain in military service worldwide with France, Cameroun, Chad, Congo, Mali, Mauritania, Niger, Senegal and Togo. A small number are civilian registered. Powered by a 450hp Pratt and Whitney Wasp radial, the Broussard cruises at 245km/h and has a maximum range of 1,200km. *Country of origin:* France. *Silhouette and picture:* MH 1521.

WSK-Okecie PZL-101A Gawron   *Confusion:* Beaver

Power: 1 × AI-14R piston engine   *Span:* 12.68m   *Length:* 9m

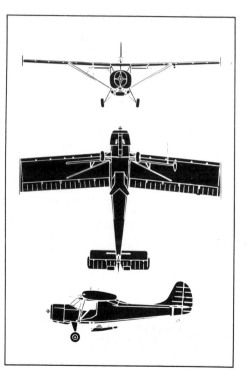

The PZL-101A Gawron (Rook) single-engined agricultural, ambulance and utility light aircraft was developed from the Russian Yak-12M and first flew in 1958. From 1961 the type was approved by a number of Eastern bloc countries as their standard agricultural aircraft, and more than 330 had been built when production ended in 1973. The Gawron has also been sold in Austria, Finland, India, Spain, Turkey and Vietnam. The aircraft is conventional in all major respects, with V-braced high wing and braced tail assembly. Powered by a 260hp AI-14R nine-cylinder radial engine, the Gawron cruises at 130km/h. *Country of origin:* Poland.

Pilatus/Fairchild PC-6 Porter

Power: 1 × Lycoming piston engine or 1 × Astazou or PT6 turboprop   *Span:* 15.14m   *Length:* 11.23m

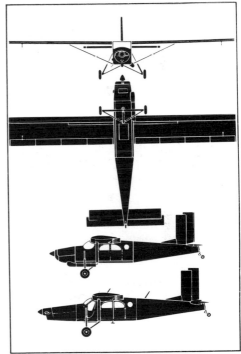

The piston-engined Pilatus Porter Stol utility aircraft was first flown in 1959. The Turbo-Porter first flew in 1961 and several variants have been built. Current production version is the PC-6/B-2-H4 Turbo-Porter, powered by a 680shp Pratt & Whitney PT6A turboprop. The type was also manufactured under licence in the USA by Fairchild Industries, the first US-built aircraft flying in 1966. A militarised version with underwing hardpoints is called the AU-23A Peacemaker. The US Army designation for the Turbo-Porter is the UV-20A Chiricahua. By early 1988 more than 470 PC-6s of all types had been ordered by operators in over 50 countries. *Country of origin:* Switzerland/USA. *Main silhouette:* Porter; *lower side view:* Turbo-Porter. *Picture:* Turbo-Porter.

PZL-Warszawa PZL-104 Wilga

Confusion: Porter

Power: 1 × Ivchenko piston engine   *Span:* 11.12m   *Length:* 8.1m

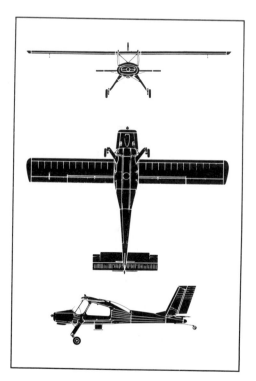

The Wilga (Thrush) was designed as a successor to the Czechoslovak L-60 Brigadyr in the agricultural role and first flew in 1962 as the Wilga 1. The design was modified in 1967 and designated the Wilga 35. Current production versions are the Wilga 35A for club flying, the Wilga 35P light transport, the agricultural 35R and the float-equipped 35H. The Wilga 80, identified by a carburettor intake mounted further aft, conforms to US requirements. More than 900 Wilgas of all versions have been built. *Country of origin:* Poland. *Silhouette:* Wilga 2. *Picture:* Wilga 35.

Confusion: Super Courier **Dornier Do 27**

Power: 1 × **Lycoming piston engine**   *Span:* 12m   *Length:* 9.6m

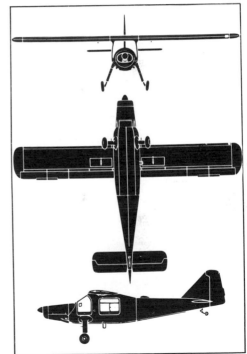

The Dornier Do 27 six/eight-seat utility aircraft first flew in 1955 in Spain, with large-scale production being undertaken in Germany from 1956. When production ended a total of 680 had been built, including 428 for the West German armed forces and 50 manufactured under licence by CASA in Spain as the CASA C.127. Several versions of the German-built aircraft have been manufactured, the principal differences between variants being engine type and internal fittings for different roles. Powered by a 340hp Lycoming, the Do 27H-2 cruises at 209km/h. Other performance figures include service ceiling of 3,300m; and a range with maximum fuel, no allowances, of 1,100km. *Country of origin:* West Germany. *Silhouette:* Do 27. *Picture:* C.127.

Power: 1 × Lycoming piston engine   *Span:* 11.89m   *Length:* 9.45m

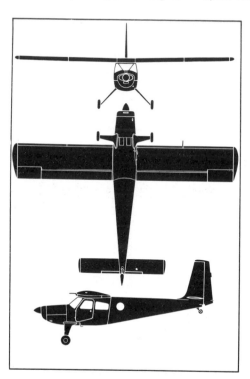

Evolved from the two-seat Koppen-Bollinger Helioplane, which first flew in 1949, the original Courier appeared in 1952. The main production version was the H-295 Super Courier, which made its maiden flight in 1958. Last production versions were the six-seat Courier 700, and the more powerful Courier 800. Both were available with wheel, amphibious float or ski undercarriages. All versions have an unbraced cantilever wing and characteristic tall fin. Powered by a 295hp Lycoming, the H-295 has a cruising speed of 241km/h. *Country of origin:* USA. *Silhouette and picture:* H-295.

Power: 1 × PT6A turboprop *Span:* 12.5m *Length:* 12m

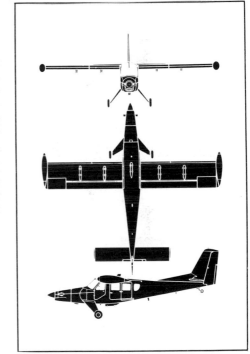

Designed as a successor to the Helio Super Courier, the turboprop Stallion first flew in 1964 and has been developed in both civil and military versions. The civil version can carry up to ten people in a high-density layout or up to six passengers in individual seats. The US Air Force acquired 15 of an armed version designated AU-24, 14 of which were subsequently supplied to the Khmer Air Force. The Stallion is easily recognisable by its swept-back main undercarriage legs and long nose accommodating the 680shp Pratt & Whitney PT6A turboprop. Cruising speed is 319km/h. A total of 600 Couriers, Super Couriers and Stallions were built. *Country of origin:* USA. *Silhouette:* AU-24. *Picture:* Model H-600B.

Atlas C4M Kudu

Confusion: Regente

Power: 1 × Lycoming piston engine   *Span:* 13.08m   *Length:* 9.31m

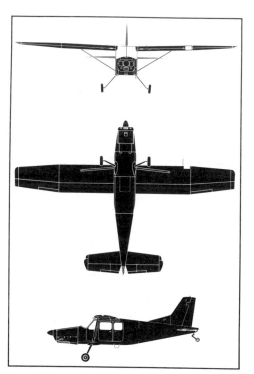

The C4M Kudu is a six/eight-seat light transport which is capable of operating from unprepared surfaces. The type first flew in 1974. The first military prototype flew in 1975 and some 35 are believed to be in service. The type is a development of the Aermacchi-Lockheed AL.60 Conestoga, which was built in both tricycle and tailwheel forms. The Kudu cruises at 200km/h and is in civil and military use in South Africa. *Country of origin:* South Africa. *Silhouette and picture:* Kudu.

Power: 1 × Continental piston engine   *Span:* 9.13m   *Length:* 7.04m

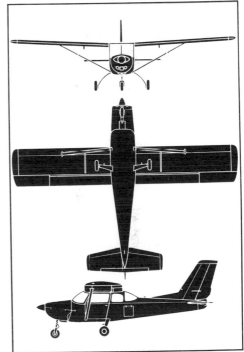

The four-seat Regente was first flown in 1961 and entered service with the Brazilian Air Force in 1963. Two versions have been built: the Regente 360C, designated C-42 in Brazilian Air Force service; and the Regente 420L, designated L-42 by the BAF. Main difference between the two is that the L-42 has a stepped-down cabin rear window for improved all-round visibility. When production ended in 1971, 80 C-42s and 40 L-42s had been built. Powered by a 210hp Continental, the L-42 cruises at 236km/h. The company also developed a civil version of the Regente, the Lanceiro, which first flew in 1970 but did not enter full-scale production. *Country of origin:* Brazil. *Silhouette and picture:* L-42.

Partenavia P.64 Oscar/P.66 Oscar/P.66C-160 Charlie

Confusion: Regente

Power: 1 × Lycoming piston engine   *Span:* 9.99m   *Length:* 7.23m

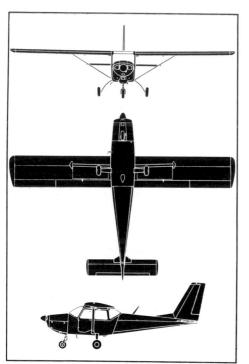

The four-seat P.64 Oscar was first flown in 1965, powered by a 180hp Lycoming engine. The following year saw the introduction of an improved version with cut-down rear fuselage decking to improve all-round visibility. A small number of aircraft were built by AFIC in South Africa as the RSA 200 Falcon. The P.66B Oscar 100 and Oscar 150 are two- and three-seat developments of the P.64. In 1976 the P-66C-160 Charlie was introduced. Powered by a 160hp Lycoming, the P-66C-160 cruises at 216km/h. *Country of origin:* Italy. *Silhouette:* P.64. *Picture:* P.66C-160 Charlie.

Cessna Model 172 Skyhawk/T-41 Mescalero

Power: 1 × Lycoming or 1 × Continental piston engine   *Span:* 10.92m   *Length:* 8.2m

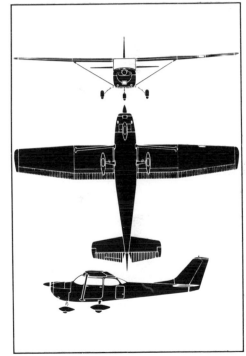

First introduced in 1955, the Model 172 was essentially a Model 170 with tricycle undercarriage and revised tail surfaces. The swept-tail 172A was introduced in 1960. Current versions are powered by a 160hp engine giving a cruising speed of about 225km/h. The Model 172E is a more powerful version with a 210hp Continental. Known in the US Army and Air Force as the T-41 Mescalero, the type is used for basic training. It has also been bought by a few South American air forces. Top of the line is the Model R-172 Hawk XP, which was also produced in France by Reims Aviation. Power is provided by a 195hp Continental. 35,773 Model 172 Skyhawks had been produced by the end of 1987, including 2,144 French-built F172s. In addition 864 military Mescaleros were built. *Country of origin:* USA. *Silhouette:* Skyhawk. *Picture:* T-41.

Cessna Model 150/152 Aerobat

Confusion: Skylane

Power: 1 × Continental or 1 × Lycoming piston engine   *Span:* 9.97m   *Length:* 7.34m

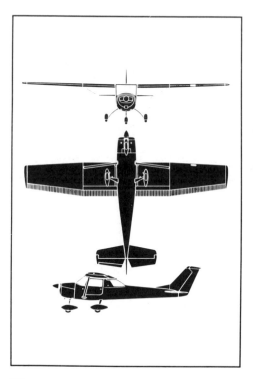

This two-seater, first flown in 1957, was intended as a successor to Cessna's earlier tailwheel Models 120 and 140. Several versions of the 150 have been built, the first major change being the introduction of the swept-tail Model 150F in 1966. Production of the basic aircraft ended in 1977 after almost 24,000 had been built, including 1,500 in France by Reims Aviation. Last production type is the Model 152, which has a more powerful engine than the Model 150 but is otherwise externally similar. The Model 152 Aerobat has certain interior modifications as well as some structural beefing-up to permit simple aerobatics. Maximum cruising speed is 188km/h. *Country of origin:* USA. *Silhouette:* Model 150F. *Picture:* Model 152.

Power: 1 × Continental or 1 × Lycoming piston engine   *Span:* 10.92m   *Length:* 8.57m

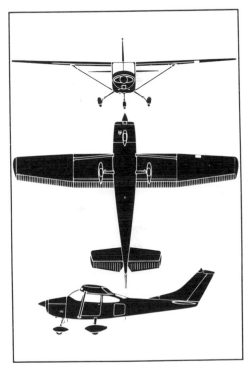

Derived from the Model 180 utility light aircraft, the Model 182 was introduced in 1956. The swept tail was introduced on 1960 aircraft. Production of the Model 182 was discontinued in 1976 and current production aircraft are the standard Skylane and Skylane II; the latter embodies a factory-installed avionics package. In 1977 Cessna introduced a new retractable-undercarriage version designated Skylane RG (and the Skylane RG II with factory-installed avionics). The RG aircraft were also fitted with the more powerful 235hp Lycoming engine. Total production of all variants exceeded 19,800, including a number built in France by Reims Aviation as the Rocket. Cruising speed is 253km/h. Retractable-undercarriage versions are known as the Skylane RG and Turbo-Skylane RG. *Country of origin:* USA. *Silhouette:* Model 182G. *Picture:* Skylane II.

Cessna Cardinal/Cardinal RG   *Confusion:* Skylane

Power: 1 × Lycoming piston engine   *Span:* 10.82m   *Length:* 8.31m

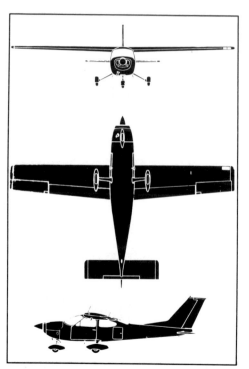

The original Model 177 four-seat light aircraft, introduced in 1967, was powered by a 150hp engine, later increased to 180hp. The 1978 aircraft were re-designated Cardinal Classic and incorporated many improvements to cabin comfort and basic equipment. The Cardinal RG series was introduced in 1970 and featured hydraulically retractable undercarriage and 200hp Lycoming engine. Cruising speed of the Cardinal Classic is 241km/h and that of the Cardinal RG 273km/h. Other performance figures for the RG include a maximum sea-level speed of 290km/h; and economical cruising speed at 3,050m of 233km/h. *Country of origin:* USA. *Silhouette:* Cardinal. *Picture:* Cardinal RG II.

Power: 1 × Continental piston engine   *Span:* 10.92m   *Length:* 8.61m

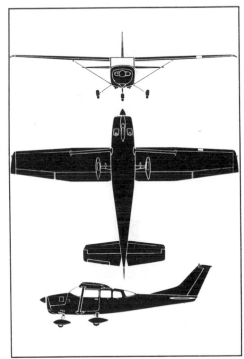

This six-seat light aircraft was introduced in 1962 as the Model 205 powered by a 285hp Continental engine. In 1964 this was superseded by the Model 206 Super Skywagon with more powerful engine and double cargo doors on the starboard side. The aircraft has since been renamed Stationair and Turbo-Stationair. 1978 aircraft were designated Stationair 6 and Turbo-Stationair 6, fitted with the 300hp Continental and 310hp turbocharged Continental respectively. The line also includes the de luxe Super Skylane, and a utility version with a single seat and no wheel fairings. In USAF service the Turbo-Stationair is designated U-26A. By the end of 1987, when production ceased, 7,652 Skywagons and Stationairs had been built. *Country of origin:* USA. *Silhouette:* Stationair. *Picture:* Stationair 6.

Cessna Stationair 7/Stationair 8

Confusion: Stationair 6, Centurion

Power: 1 × Continental piston engine   *Span:* 10.92m   *Length:* 9.68m

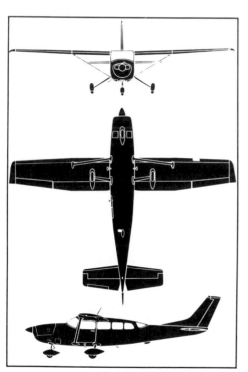

This seven-seat utility aircraft, first announced in 1969, is a stretched version of the Model 206 Super Skywagon. In addition to the longer fuselage, new features included a door for the co-pilot on the starboard side and a separate baggage compartment forward of the cabin. The basic aircraft is powered by a 300hp Continental and cruises at 265km/h, while the turbocharged aircraft with a 310hp Continental cruises at 297km/h. Production of Model 207s had exceeded 360 by early 1977. The designation was changed to Stationair 7 and Turbo-Stationair 7 for the 1978 aircraft. In 1980 production switched to the eight-seat, slightly longer Stationair 8/Turbo Stationair 8. *Country of origin:* USA. *Silhouette:* Skywagon 207. *Picture:* Stationair 8.

Power: 1 × PT6A turboprop   *Span:* 15.8m   *Length:* 11.46m

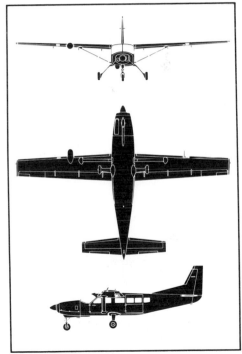

The turboprop Model 208 Caravan 1 was a completely new Cessna design first flown in December 1982. Intended to serve as a transport from unprepared strips, the Caravan can carry up to 13 passengers. Wheel, ski and float undercarriages are available and the Caravan is also designed to perform a variety of military and civilian utility duties. A military derivative is the U-27A. The freighter version, 208A, carries a large underfuselage pannier and has no windows. A stretched freighter is the 208B. By early 1988 193 Caravan 1s had been delivered. Maximum cruising speed is 342km/h. *Country of origin:* USA. *Silhouette:* Caravan 1. *Picture:* UH-27A.

Cessna Model 210/Centurion

Confusion: Stationair 6/7

Power: 1 × Continental piston engine   *Span:* 11.2m   *Length:* 8.59m

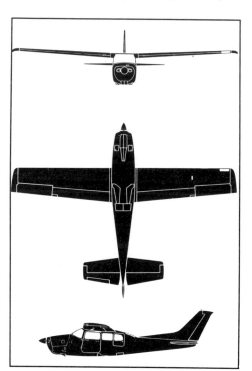

The original Model 210 first flew in 1957 and was the first of the Cessna high-wing range of light aircraft to feature a retractable undercarriage. Later versions also featured an unbraced cantilever wing, as on the Cardinal series. Successive models have introduced enlarged cabin windows. The T210 is now known as Turbo Centurion with two additional rear windows and increased cabin headroom. In late 1977 Cessna introduced the Pressurised Centurion, generally similar to the standard aircraft but powered by a 310hp Continental engine and with four small windows on each side in place of the panoramic windows. Maximum cruising speed is 303km/h. *Country of origin:* USA. *Silhouette:* Model 210H. *Picture:* Pressurised Centurion.

Confusion: Safari

MFI MFI-9B Trainer/Bölkow BO 208 Junior

Power: 1 × Continental piston engine   *Span:* 7.43m   *Length:* 5.85m

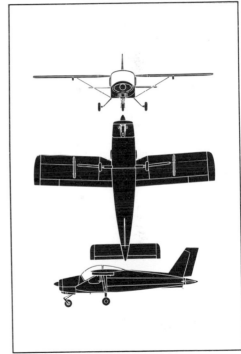

The MFI-9B Trainer and MFI-9 Junior are both derivatives of the BA-7 built in the USA by Bjorn Andreasson and first flown in 1958. A total of 25 were built by MFI at Malmo in Sweden. In 1962 Bölkow in West Germany began producing the type under licence from MFI, and by late 1969 production had reached 200 aircraft. Fully aerobatic, the Junior seats two people side-by-side and is powered by a 100hp Rolls-Royce Continental. Other performance figures include a maximum speed of 225km/h; economical cruising speed of 195km/h; and a range of 748km. Certified in the normal, utility and glider-towing categories, the MFI-9 is of all-metal construction. The number of Juniors remaining airworthy is dwindling steadily. *Country of origin:* Sweden/West Germany. *Silhouette:* MFI-9B. *Picture:* BO 208.

Saab/Safari/Supporter

Confusion: Junior

Power: 1 × Lycoming piston engine   *Span:* 8.85m   *Length:* 7m

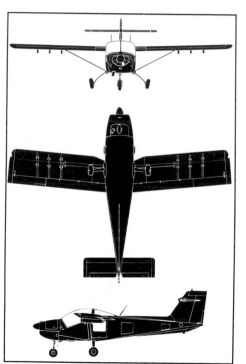

The two-seat Saab Safari (formerly known as the MFI-15) basic trainer and utility aircraft first flew in 1969, powered by a 160hp engine. Originally flown with a low-set tailplane, the type was refitted with a near T-tail. Although designed with a tricycle undercarriage, the type may be seen with an optional tailwheel arrangement. A military version is also available, known as the Supporter. Externally similar to the Safari, the Supporter is stressed to carry stores on six underwing stations. Supporters have been delivered to Denmark, Pakistan and Zambia. Powered by a 200hp Lycoming, both versions cruise at 210km/h. *Country of origin:* Sweden. *Silhouette and picture:* Supporter.

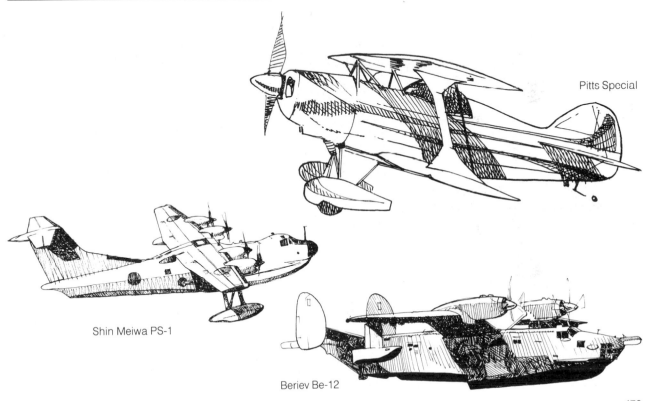

Pitts Special

Shin Meiwa PS-1

Beriev Be-12

Lake LA4 Buccaneer *Confusion:* Teal

Power: 1 × Lycoming piston engine   *Span:* 11.58m   *Length:* 7.6m

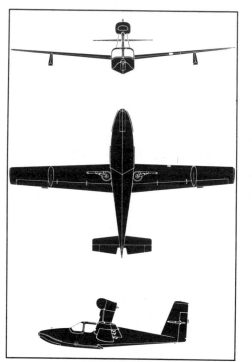

The LA4 Buccaneer four-seat light amphibian is the production development of the C-1 Skimmer, which first flew in 1948. The first Lake-built aircraft, designated LA4, flew in 1960. Two versions are currently built, the standard LA4-200 EP, and the LA4-200 EPR with a reversible pitch propeller. A six-seat development, the LA-250 Renegade, with a longer fuselage and more powerful engine, is now in production. Over 1,200 LA4s of all versions have been built. Powered by a 200hp Lycoming, the standard LA4-200 has a cruising speed of 248km/h and range with maximum fuel at maximum cruising speed is 1,046km. *Country of origin:* USA. *Silhouette and picture:* LA4-200.

Power: 2 × PT6A turboprops   *Span:* 15.5m   *Length:* 12.46m

The Claudius Dornier company (not to be confused with the Dornier company) has produced a 12 passenger twin turboprop amphibian of which initial deliveries were due to commence in mid 1990. The prototype, later damaged, made its first flight in August 1984 and was followed, in developed form, by a second machine in April 1987. The pusher/tractor propeller engines are mounted in tandem above the parasol wing which has a marked similarity in planform, to that of the Dornier 228. Production CD2s will have largely glassfibre hulls and carbon fibre wings. A variety of missions, civil and military, are envisaged apart from purely passenger transport. Maximum cruising speed of the CD2 is 341km/h and range 555km. *Country of origin:* West Germany.

Schweizer TSC-1 Teal

Confusion: Lake Buccaneer

Power: 1 × Lycoming piston engine   *Span:* 9.73m   *Length:* 7.19m

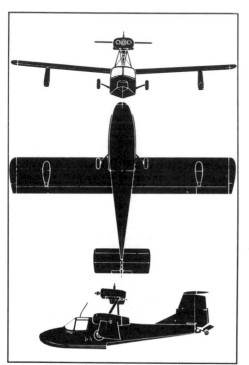

The two-seat all-metal Teal amphibian was developed originally by Thurston Aircraft and first flown in 1968. Design and production rights were acquired by Schweizer in 1972. Production of the Teal I ran to 15 aircraft. The Teal II incorporates several minor design changes. The entire Teal programme was transferred to Teal Aircraft Corporation at St Augustine, Florida, in the spring of 1976. Last known production versions were the Teal II and the improved Teal III, which has a more powerful Lycoming engine and can accommodate up to four people. The standard Teal II cruises at 185km/h. *Country of origin:* USA. *Silhouette and picture:* Teal II.

Confusion: Super Widgeon # McKinnon G-21 Turbo-Goose

Power: 2 × PT6A turboprops　　*Span:* 15.49m　　*Length:* 12.07m

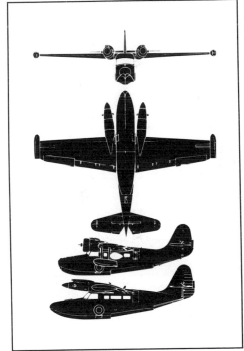

The original six/seven-seat Grumman G-21 first flew in 1937, powered by a pair of 450hp Wasp Junior radials. In the conversion, carried out by McKinnon Enterprises of Oregon, the radial engines are replaced by a pair of 680shp turboprops giving a maximum speed of about 354km/h. Other changes include the fitting of retractable wingtip floats, longer nose, dorsal fin and larger cabin windows as well as numerous interior and equipment changes. The first Turbo-Goose conversion was flown in 1966. There is accommodation for 9 to 12 people. The company subsequently became McKinnon-Viking and moved to Canada. Conversion work has ceased. *Country of origin:* USA. *Main silhouette:* Turbo-Goose; *upper side view:* G-21. *Picture:* Turbo-Goose.

McKinnon Super Widgeon *Confusion:* Turbo-Goose

Power: 2 × Lycoming piston engines *Span:* 12.9m *Length:* 9.47m

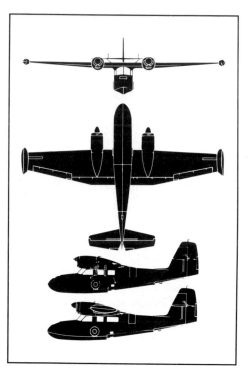

The original G-44 Widgeon was a four-seat light amphibian and first flew in 1940. Although designed as commercial transports the 176 Widgeons built during the Second World War were used as three-seat anti-submarine patrol aircraft. Subsequent civil aircraft were five-seaters, and a number were built in France as the SCAN 30. The McKinnon Super Widgeon is an executive conversion with the original Ranger in-line engines replaced by two 270hp Lycomings, giving a cruising speed of 290km/h. The company converted more than 70 aircraft, a number of which have retractable wingtip floats. Other features include picture windows, a modern instrument panel, improved soundproofing and an emergency escape hatch. Three or four passengers can be carried over a maximum-fuel range of 1,600km. *Country of origin:* USA. *Main silhouette:* Super Widgeon; *upper side view:* G-44. *Picture:* Super Widgeon.

Power: 2 × R-1820 piston engines   *Span:* 29.46m   *Length:* 19.38m

Designed for a US Navy requirement for a utility transport and air-sea rescue amphibian, the HU-16 Albatross first flew in 1947. It was subsequently selected by the US Air Force and a total of more than 300 were built in several versions, including the HU-16D for the US Navy and the HU-16E for the US Coast Guard. Small numbers remain in service with Greece, Malaysia, Mexico, the Philippines and Taiwan, and 12 aircraft have been converted by Grumman Aerospace to 28-seat configuration for civil use under the designation G-111. Typical cruising speed of the G-111 is 382km/h. *Country of origin:* USA. *Silhouette and picture:* HU-16B.

485

Canadair CL-215 *Confusion:* Albatross

Power: 2 × R-2800 piston engines   *Span:* 28.6m   *Length:* 19.82m

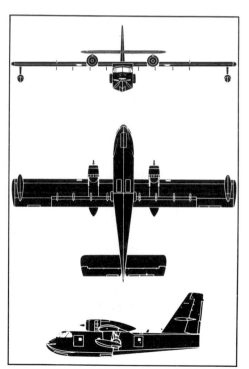

The Canadair CL-215 amphibian first flew in October 1967. Designed primarily for firefighting, the CL-215 serves also in air-sea rescue, coastal patrol and passenger transport roles. For fire-fighting a load of 5,346 litres of water can be carried, and scoops in the hull of CL-215 enable it to pick up a full load from the surface of a lake or ocean in 10 seconds. More than 100 aircraft had been ordered by 1985 by operators in Canada, France, Spain, Greece, Italy, Thailand, Venezuela and Yugoslavia. Powered by two Pratt and Whitney R-1800 radial engines, the CL-215 cruises at 291km/h. The latest version is the CL-215T converted to turboprop power with PW123AF engines. The CL-215T has upturned winglets at the tips. *Country of origin:* Canada.

Power: 2 × Ivchenko turboprops   *Span:* 29.7m   *Length:* 32.9m

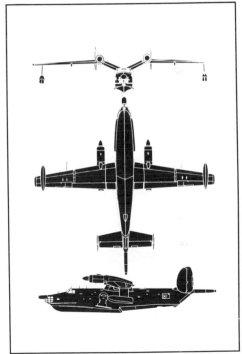

The twin-engined Be-12 maritime patrol amphibian is in current service with the Soviet Naval Air Force. A successor to the Be-6, it was displayed publicly for the first time at Moscow Tushino in 1961. The Mail has a normal operating speed of about 322km/h. Armament is carried in an internal bomb bay and there is provision for two stores pylons underneath each wing. About 100 Be-12s are thought to have been built. The Be-12 holds every one of the FAI world records in its class, no fewer than 38 in all. The Soviet Union has developed a new amphibian, the A-40 'Tag-D', as a replacement for the Be-12. *Country of origin:* USSR.

487

Shin Meiwa PS-1/US-1   *Confusion:* Mail

Power: 4 × T64 turboprops   *Span:* 33.15m   *Length:* 33.46m

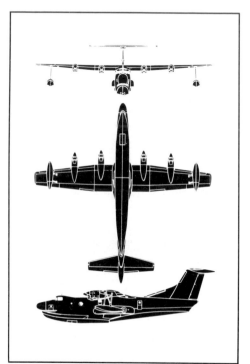

Developed to fill a Japanese Maritime Self-Defence Force requirement for an anti-submarine flying boat (PS-1) and a search and rescue amphibian (US-1), this aircraft, which carries the company designation SS-2 (SS-2A for the amphibious version), first flew in October 1967. Some 23 PS-1s, including two prototypes, were delivered. The first US-1 flew in 1974 and ten are now in service. Later aircraft in this series are fitted with more powerful engines and are designated US-1A. Cruising speed of the US-1A is 426km/h. *Country of origin:* Japan. *Silhouette and picture:* SS-2A.

Power: 4 × Harbin turboprops   *Span:* 36m   *Length:* 38.9m

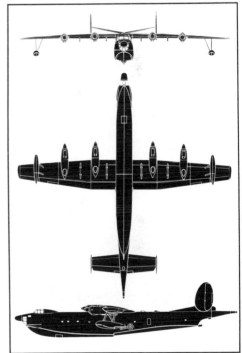

First flown in April 1976, the SH-5 flying boat amphibian represents a rare breed in current designs. It fulfills a variety of roles including maritime patrol/ASW, surveillance, search and rescue and water bombing. The type entered service in 1986 following a protracted development programme. The SH-5 normally carries a crew of eight. The nose is equipped with a search radar, while there is a magnetic anomaly detector fitted in the tail. Maximum all-up weight is 26,500kg, maximum cruising speed 450km/h and maximum range 4,750km. *Country of origin:* China.

de Havilland D.H.82 Tiger Moth *Confusion:* SV.4

Power: 1 × de Havilland piston engine   *Span:* 8.94m   *Length:* 7.29m

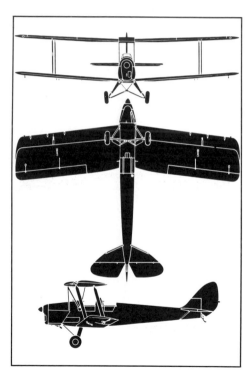

Probably the most famous training aircraft of all, the Tiger Moth first flew in 1931. When production ended more than 7,000 had been built in Britain, Canada, Australia and New Zealand. A number are still airworthy. Its distinctive de Havilland-style tail and swept-back wings make the Tiger Moth fairly easy to recognise. Powered by a 130hp Gipsy Major engine, the Tiger Moth cruises at 145km/h over a range of 483km. *Country of origin:* UK.

Power: 1 × Renault or 1 × de Havilland piston engine   *Span:* 8.38m   *Length:* 6.96m

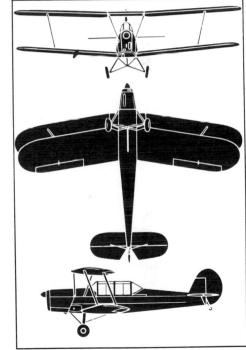

The SV.4 tandem two-seat light biplane was first flown in 1933, powered by a 130hp Gipsy III engine. Used originally as a primary trainer by the Belgian Air Force, it has since been much sought after as a sporting aircraft by virtue of its excellent aerobatic qualities. Before the advent of the Pitts and Zlin the Stampe SV.4 was the mainstay of British competition aerobatics. Bearing a superficial resemblance to the Tiger Moth, the Stampe may be recognised by its rounded fin and rudder, rounded wingtips, and ailerons on both upper and lower wings. Powered by the 130hp Gipsy Major, the SV.4 cruises at 177km/h. *Country of origin:* Belgium.

Power: 1 × Lycoming piston engine   *Span:* 8.13m   *Length:* 6.2m

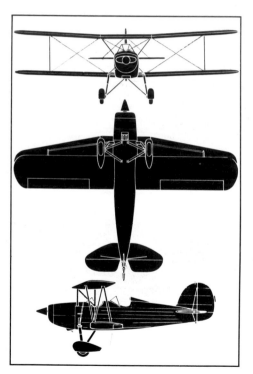

The original Model 2T-1 tandem two-seat light biplane was built in substantial quantities by the Great Lakes Aircraft Co between 1929 and 1932 and was fitted with a variety of engines ranging in power from 95 to 200hp. The company was taken over in 1972 and subsequently the Great Lakes Sports Trainer was put back into production, initially powered by a 140hp Lycoming. Following suspension of production from 1982, the line was restarted in 1984. Current aircraft are fitted with 180hp engines. *Country of origin:* USA. *Silhouette and picture:* Sports Trainer.

Power: 1 × Lycoming piston engine *Span:* 5.28m *Length:* 4.71m

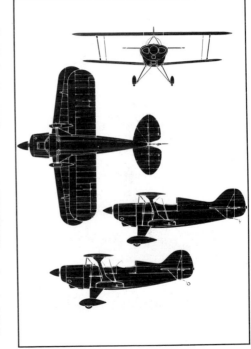

First flown in single-seat form in 1944, the Pitts Special has become the best-known competition aerobatics aircraft. Now built by Christen Industries, both single seat S-1S/S-2S and two-seat S-2A/S-2B versions are manufactured. An advanced single-seat version known as the S-1T special is available. All versions can be purchased in kit form. Powered by a 180hp Lycoming engine, top speed of the S-1S is 283km/h. *Country of origin:* USA. *Main silhouette:* S-1S; *lower side view:* S-2A. *Picture:* S-1S.

 **Bücker Jungmann/Jungmeister**   *Confusion:* Pitts Special

Power: 1 × Hirth (Jungmann) or   *Span:* (Jungmann): 7.4m   *Length:* (Jungmann): 6.6m
1 × Siemens piston engine (Jungmeister)

The tandem two-seat Jungmann was first flown in 1934 and several remain in use in various parts of the world. The type has been licence-built in several countries, including Czechoslovakia, Switzerland and Spain, with production in the latter country resuming in 1956 with the manufacture of 55 aircraft by CASA. The single-seat Jungmeister was introduced in 1935 as an advanced aerobatic biplane and, like the Jungmann, the type was licence-built in Spain and Switzerland. Limited production was restarted in 1968 by Aero Technik Canary in West Germany, the aircraft being standard apart from the wheels and instruments. *Country of origin:* West Germany. *Main silhouette:* Jungmeister; *lower side view:* Jungmann. *Picture:* Jungmeister.

Power: 1 × R-1820 piston engine   *Span:* 12.7m   *Length:* 9.14m

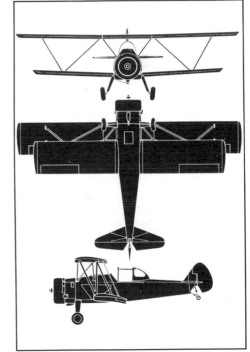

The first Emair MA-1 was built by Air New Zealand engineers under contract to a Hawaiian company and first flew in 1969. Subsequently 25 were produced in Texas by Emair. Production of the MA-1 ended in 1976 and a new version, the MA-1B Diablo 1200, was put into production, 48 being built before the line was closed in 1980. Powerplant is a 1,200hp Wright R-1820 radial derated to 900hp, giving a cruising speed of 185km/h. As in most agricultural aircraft, the cockpit is raised well above the level of the fuselage top decking and incorporates a hefty roll-bar. The chemical hopper is built into the front fuselage and has two outlets. *Country of origin:* USA. *Silhouette:* MA-1. *Picture:* MA-1B.

Power: 1 × R-985 or 1 × R-1340 piston engine   *Span:* 10.95m   *Length:* 7.42m

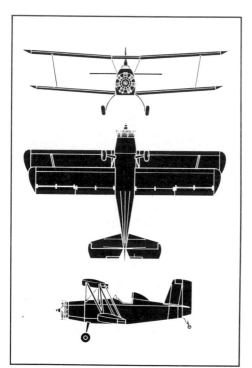

The original Ag-Cat agricultural biplane first flew in 1957 and was manufactured by Schweizer under licence from Grumman; Schweizer has since taken over all rights to the Ag-Cat. The Super Ag-Cat was first certificated in 1966 and was fitted with a 600hp Pratt & Whitney. Several versions of the Super Ag-Cat have been built, with a variety of power options; the B and C versions are also slightly larger. The Turbo Ag-Cat D has a very long nose housing a Pratt & Whitney PT6A turboprop. All Ag-Cats have the distinctive square fin and rudder and are considerably shorter than the other US-built agricultural biplane, the Emair MA-1B. *Country of origin:* USA. *Silhouette:* Ag-Cat. *Picture:* Super Ag-Cat.

Power: 1 × Shvetsov radial   *Span:* 18.18m   *Length:* 12.74m

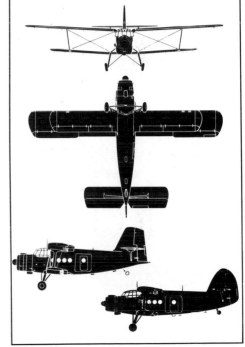

The first An-2 made its maiden flight in 1947. No longer produced in the Soviet Union, the An-2 is manufactured in Poland by the WSK factory at Mielec. Others have been built in China under the designation Y-5, and total production exceeds 16,000. The An-2 is used for a variety of tasks including crop spraying, transport and photography. Powered by a nine-cylinder radial, the An-2 has a cruising speed of 185km/h. *Country of origin:* Poland/USSR. *Main silhouette:* An-2; *upper side view:* An-2M. *Picture:* An-2R.

de Havilland D.H.100/Vampire

Confusion: Skymaster

Power: 1 × Goblin turbojet   *Span:* 11.6m   *Length:* 9.37m

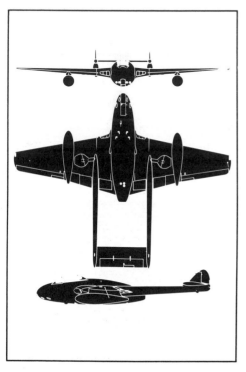

First flown in 1943, the Goblin-powered Vampire fighter was built in large numbers for more than 14 years, but only a handful remain in service. Switzerland has 35 T55 two-seat trainers and about 20 FB6s. The Vampire is easy to recognise with its short fuselage and twin booms. Typical maximum speed of the Vampire is 870km/hr and ceiling is 12,000m. *Country of origin:* UK. *Silhouette and picture:* Vampire FB6.

Power: 2 × Continental piston engines   *Span:* 11.63m   *Length* 9.07m

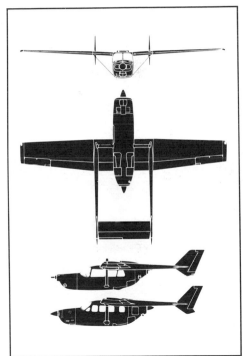

This all-metal four/six-seater is unusual in having a 210hp Continental mounted at each end of the fuselage nacelle. First flown in 1961, it entered series production as the fixed-undercarriage Model 336 Skymaster in May 1963. A military version, designated O-2, has also been delivered to the USAF and Iranian Air Force for forward air controller and psychological warfare missions. Military deliveries totalled 513. The Model P337 is a pressurised version distinguished by having four instead of three fuselage cabin windows. Deliveries of all variants had totalled 2,993 by the time production was terminated in 1980. Performance figures for the P337 include a maximum level speed of 402km/h, maximum cruising speed of 380km/h, and a range at economical cruising speed of 2,140km. *Country of origin:* USA. *Main silhouette:* Model P337; *upper side view:* O-2. *Picture:* Model P337.

IAI 101/102/201 Arava

Confusion: Noratlas

Power: 2 × PT6A turboprops   *Span:* 20.96m   *Length:* 13.03m

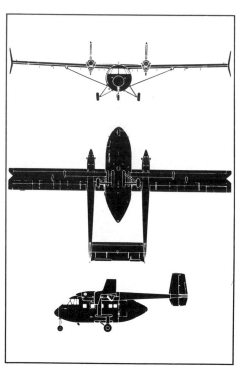

The Arava was designed to fulfil a requirement for a light Stol transport for civil and military use. First flight was in late 1969. Four versions have been announced to date. Three of these are externally similar: the IAI 101, the first civil version; the IAI 102, the second civil version, based on the original 101 and able to accommodate up to 20 passengers; and the IAI 201, the military transport version. Based on the original 101, it first flew in 1972. Latest variant is the IAI 202, distinguished by its wingtip winglets. Electronic-warfare versions of the IAI 201 are in service with the Israeli Air Force. Production of the Arava finished in 1987. *Country of origin:* Israel. *Silhouette:* IAI 202. *Picture:* IAI 201.

Power: 2 × Hercules piston engines *Span:* 32.5m *Length:* 21.96m

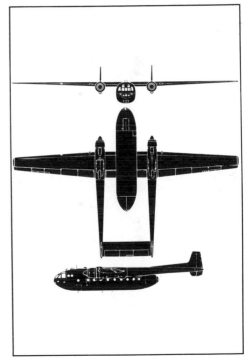

The twin-engined, twin-boom Noratlas was the standard medium tactical transport of both the French and West German air forces, which received 211 and 188 respectively. Small numbers of aircraft serve with six air forces round the world. Five crew and up to 45 troops can be carried, or alternatively, up to 6,803kg of freight. Similar in size to the Fairchild C-119 Flying Boxcar, the Noratlas is identified by squared-off rudders, narrower wings and a distinctive stepless nose. Powered by two 2,090hp Snecma-built Bristol Hercules radials, the Noratlas cruises at 322km/h. *Country of origin:* France.

Fairchild C-119 Flying Boxcar

Confusion: Noratlas, Arava

Power: 2 × R-3350 piston engines   *Span:* 33.3m   *Length:* 27.25m

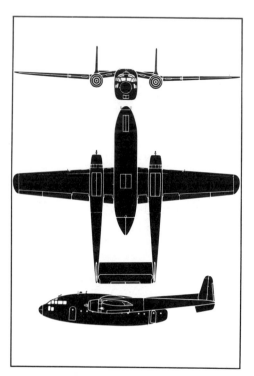

The C-119 is a development of the C-82 Packet, which first flew towards the end of the Second World War. Retaining the twin-engine, twin-boom arrangement of the Packet, the Flying Boxcar has a longer and slimmer fuselage. When production ended in 1955 some 1,100 C-119s had been built. Although no longer used in quantity by US forces, the type is still in service with Taiwan. Powered by two 3,400hp Wright R-3350 radials, the C-119 cruises at 320km/h. Maximum payload is 9,070kg, which can be carried over a range of 1,595km. Ferry range is 5,570km. *Country of origin:* USA.

Power: 2 × T76 turboprops   *Span:* 12.19m   *Length:* 12.67m

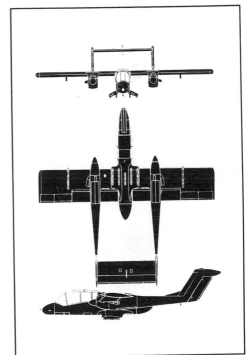

The Bronco was North America's winning entry in the US Navy's Light Armed Reconnaissance Aircraft (LARA) competition and the first prototype flew for the first time in mid-1965. The OV-10A, the basic production aircraft, is in service with the USAF, and US Marine Corps. This variant has also been delivered to four other countries. The OV-10B is used for target towing, while the OV-10B(Z) is equipped with an auxiliary turbojet mounted above the fuselage nacelle. The OV-10D is the Night Observation Gunship System (NOGS) with increased armament, including a 20mm gun turret beneath the rear fuselage. *Country of origin:* USA. *Silhouette:* OV-10A. *Picture:* OV-10D.

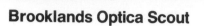

Brooklands Optica Scout

Confusion: Bronco

Power: 1 × Lycoming piston engine and fan   *Span:* 12m   *Length:* 8.15m

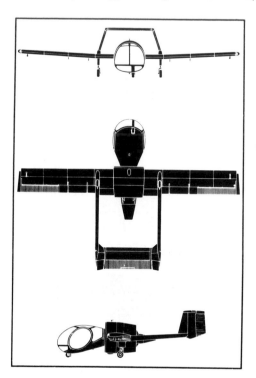

Like the Lear Fan, the Optica Scout is not a jet but it looks like one in the air. A unique design, this three-seater with twin booms and high tail is designed for aerial surveillance of all types and photography. It can loiter at speeds as low as 98km/h and is much cheaper to operate than a helicopter. The Lycoming piston engine drives a fixed-pitch fan; this combination is known as a ducted propulsion unit. First flown in December 1979, the Optica was originally known as the Edgley EA7 Optica. The Edgley Company was renamed Optica Industries and, later, Brooklands Aircraft Ltd. The firm is aiming to produce one Optica Scout per month. *Country of origin:* UK.

PZL-Mielec M-15 Belphegor

Power: 1 × Ivchenko turbofan   *Span:* 22.4m   *Length:* 12.72m

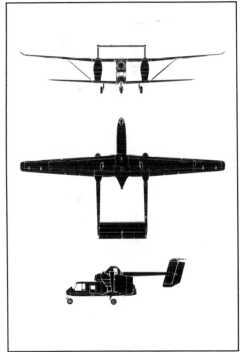

Produced as a result of an anticipated Soviet requirement for up to 3,000 large agricultural aircraft, the Polish-built M-15 Belphegor first flew in 1973. The full-chord interplane struts are used as streamlined chemical hoppers, while the 1,499.4kg-thrust turbofan is mounted in a pod on top of the fuselage. The single pilot is accommodated in the extreme nose of the aircraft and there is a cabin to the rear of the cockpit for two extra people during ferry flights. A tandem two-seat trainer version has been built. Production ended in 1981 after 180 aircraft had been constructed. *Country of origin:* Poland.

505

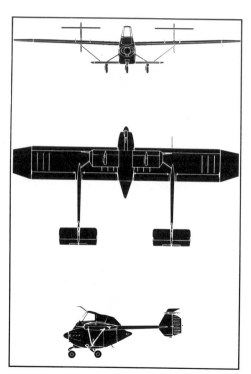

Transavia PL-12/T-320 Airtruk/Skyfarmer

Confusion: ——

Power: 1 × Continental piston engine or 1 × Textron Lycoming turboprop   *Span:* 11.98m   *Length:* 6.35m

This single-engined agricultural aircraft has one of the most distinctive layouts imaginable, the twin-boom arrangement having been adopted to keep the tails clear of chemicals and also to permit rapid loading from a vehicle. First flown in 1966, the type has been exported to nine countries. A utility version designated the PL-12-U has also been built. This variant can carry one passenger on the upper deck behind the pilot's position and another four on the lower deck. The T-320 Airtruk differs principally in having a 325hp Continental Tiara in place of the PL-12's 300hp Rolls-Royce Continental. A development is the T-300 Skyfarmer with an Avco Lycoming engine. Skyfarmer T-300A has a larger upper fuselage, while T-400 has a higher power turboprop. *Country of origin:* Australia. *Silhouette and picture:* PL-12.

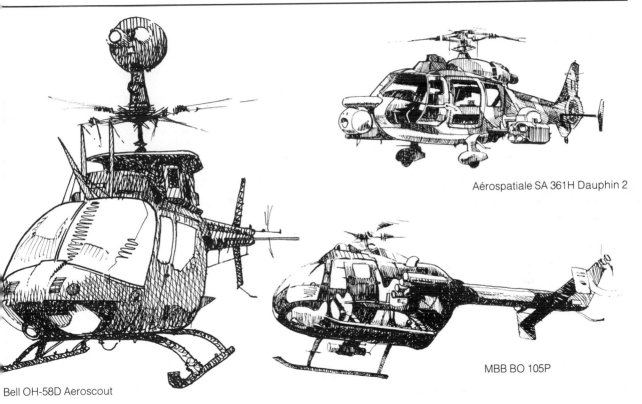

Aérospatiale SA 361H Dauphin 2

MBB BO 105P

Bell OH-58D Aeroscout

Rogerson Hiller UH-12/Model 360/H-23 Raven   *Confusion:* Bell 47, Alouette II, R22

Power: 1 × Franklin or 1 × Lycoming piston engine   *Rotor dia:* 10.8m   *Length:* 8.69m

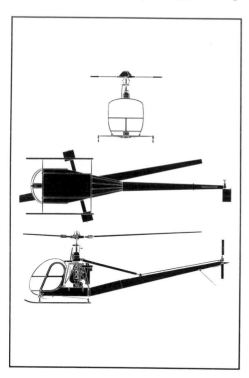

The UH-12 was first produced in 1948 and was followed in 1950 by the UH-12A with a semi-enclosed cockpit. The UH-12B was the military variant and the UH-12C introduced the "goldfish bowl" cabin. The Models A, B and C were all powered by 200hp or 210hp Franklin engines. The Model 12E was powered by the 305hp Lycoming engine. Hiller Aviation was reformed in 1973 and was taken over by Rogerson Aircraft in 1984, the current designation being UH-12E Hauler for the basic three-seat version. The H-23 Raven is a military version of the Hiller 12. Performance figures for the Model 12E/Raven include a maximum speed of 154km/h, cruising speed of 145km/h and sea-level range of 362km. Deliveries of the UH-12 were resumed in 1984. *Country of origin:* USA. *Silhouette and picture:* Hiller 12E.

Power: 1 × Lycoming piston engine   *Rotor dia:* 11.32m   *Length:* 9.63m

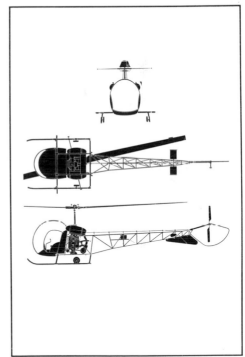

The basic Bell Model 47 first flew in 1945, and the first production models were ordered for service with the US Army and Navy. Many versions have subsequently been built for military service in a variety of countries, and a large number used by civil operators. The Model 47 was also built under licence in Italy as the Agusta-Bell 47 series, and in the UK by Westland as the Sioux AH.1. All models from the 47D onwards have the familiar "goldfish bowl" cabin and open tail-boom. Subsequent variants had engines of differing powers, but the main design change came with the Model 47H and Model 47J Ranger, which had a fully enclosed fuselage and cabin. These changes apply also to Agusta-built aircraft. *Countries of origin:* Italy/UK/USA. *Silhouette:* Bell 47G2. *Picture:* Bell 47G3 Sioux.

 **Robinson R22 Alpha** *Confusion:* Raven

Power: 1 × Lycoming piston engine   *Rotor dia:* 7.67m   *Length:* 8.76m

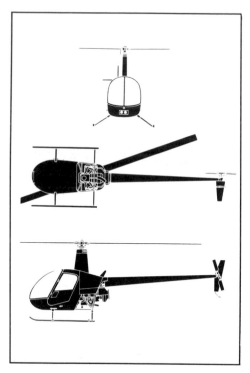

Over 800 Robinson R22 light utility helicopters have been built, largely for civil pilot training. One of the cheapest two-seaters on the market, the R22 has a top speed of 187km/h and cruises at 174km/h. Maximum range is 386km. The prototype R22 first flew in 1975. The current production model, the R22 Beta, has a higher-powered Lycoming 0-320-B2C engine. With floats it is known as the R22 Mariner *Country of origin:* USA.

Power: 1 × Artouste or 1 × Astazou turboshaft   *Rotor dia:* 11.02m   *Length:* 10.26m

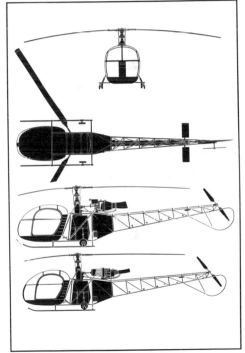

All the helicopters in this family are basically similar and differ principally in their powerplants. The prototypes were built as SE.3120s and powered by a Salmson 9 engine; production types were designated SE.3130 and subsequently SE.313B Alouette II. Production totalled more than 900. The Astazou-powered SA.318C first flew in 1961 and when production ended in 1975 more than 350 had been built. The SA.315B Lama is an Artouste-powered version built to an Indian requirement, while the Cheetah is a licence-built version produced by HAL for the Indian Army. The Lama has been sold to operators in more than 20 countries. *Country of origin:* France. *Main silhouette:* SA.318C; *lower side view:* SE.313B. *Picture:* SA.318C.

Aérospatiale Alouette III

Confusion: Alouette II, Gazelle

Power: 1 × Artouste or 1 × Astazou turboshaft   *Rotor dia:* 11.02m   *Length:* 10.03m

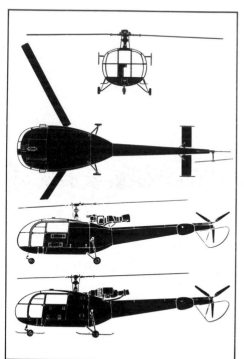

The Alouette III series was derived from the Alouette II, offering a larger cabin, more powerful engine and improved performance. The type first flew in 1959 as the SE.3160, and later production aircraft were designated SA.316B. Some 1,455 Alouette IIIs of all versions had been delivered to a number of operators in more than 70 countries by the time production ceased in 1985. The SA.316B has also been built under licence in India as the Chetak, and in both Switzerland and Romania. The SA.319B, a direct development of the SA.316B, is powered by a 870shp Turboméca Astazou engine. First flown in 1967, the type is still in production in Romania. Powered by an 870shp Turboméca Artouste turboshaft, the SA.316B cruises at 185km/h, while the SA.319B cruises at about 190km/h. *Country of origin:* France. *Main silhouette:* SA.316B; *lower side view:* SA.319B.

Aérospatiale SA.341/342 Gazelle

Power: 1 × Astazou turboshaft   *Rotor dia:* 10.5m   *Length:* 11.97m

This five-seat light helicopter first flew in 1967, powered by an Astazou III engine. The first production aircraft flew in 1971, differing from the prototype in having a longer cabin, larger tail unit and uprated Astazou IIIA. Eleven versions of the Gazelle have been built to date: the SA.341B, 341C, 341D and 341E for the British armed forces; the SA.341F for the French Army; the SA.341G and SA.342J for civil operators; SA.341H military export version; and the SA.342K, 342L and 342M military versions. Under an Anglo-French agreement signed in 1967, Gazelles were produced jointly with Westland Helicopters and are also built under licence in Yugoslavia and Egypt. Total sales have exceeded 1,400. *Country of origin:* France. *Silhouette:* SA.341. *Picture:* Gazelle HT.2.

Aérospatiale SA.360 and SA.361H Dauphin *Confusion:* Gazelle

Power: 1 × Astazou turboshaft (SA.360 and SA.361) or   *Rotor dia:* 11.5m   *Length:* 13.2m
       2 × Arriel turboshafts (SA.365)

Developed as a replacement for the Alouette III, the SA.360 Dauphin first flew in 1972 powered by a 980shp Turboméca Astazou XVI turboshaft. Production aircraft are powered by the 1,050shp Astazou XVIIIA. The SA.361H Dauphin is similar in most respects to the SA.360 but is powered by a 1,400shp Astazou XXB. The SA.365C Dauphin 2 is marginally larger than the standard Dauphin and differs principally in having two 650shp Turboméca Arriel engines and accommodation for up to 13 persons. Cruising speed for both versions is 275km/h. *Country of origin:* France. *Silhouette:* SA.360. *Picture:* SA.360

Power: 2 × Arriel turboshafts   *Rotor dia:* 11.93m   *Length:* 11.44m

Confusing because it is also called Dauphin 2, the 365N is in fact a major redesign compared with the 365C, with extensive use of composite materials and a retractable undercarriage. The 365N is a 10–14-seat transport, 365M a military utility vehicle and 365F a maritime version for anti-shipping and search and rescue. The SA.366, with two Avco Lycoming turboshafts, has been ordered by the US Coast Guard for short-range recovery. In this form it has the US designation HH-65A Dolphin. A multi-role military development is the SA.365M Panther. The 365N is being produced under licence in China and several hundred of the variants have been built. *Country of origin:* France. *Silhouette and picture:* SA.365F.

Westland Scout/Wasp *Confusion:* Bell 204

Power: 1 × Nimbus turboshaft   *Rotor dia:* 9.83m   *Length:* 9.24m

Derived from the Saunders-Roe P.531, which first flew in 1958, the Scout army helicopter is fitted with a skid undercarriage. The British Army took delivery of about 150 Scouts of which 30 are still operational. The Wasp naval version, distinguished by its wheeled undercarriage, was first delivered to the Royal Navy in 1963. Wasps are used in Brazil, New Zealand, Indonesia, Malaysia and South Africa. Powered by a 710shp Rolls-Royce Nimbus turboshaft, the Scout cruises at about 210km/h and the Wasp at 175km/h. *Country of origin:* UK. *Main silhouette:* Scout; *lower side view:* **Wasp**. *Picture:* Scout.

Power: 1 × T53 turboshaft   *Rotor dia:* 13.41m   *Length:* 11.7m

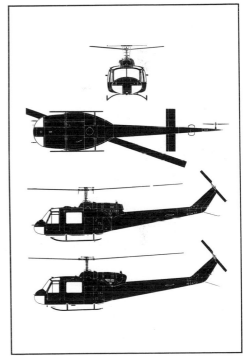

The Model 204 and UH-1 Iroquois belong to a family of commercial utility and military general-purpose helicopters derived from the XH-40, which first flew in 1956. The civil and military versions are essentially similar in external appearance and differ principally in their internal equipment. A licence-built version, the Agusta-Bell 204B, was in production in Italy between 1961 and 1974. Italian-built aircraft were fitted with the Rolls-Royce Gnome turboshaft. *Country of origin:* USA/Italy. *Main silhouette:* Bell 204; *lower side view:* Agusta-Bell 204B. *Picture:* Agusta-Bell 204B.

Bell Model 212/412 Model 212ASW *Confusion:* Bell 204, Bell 205

Power: 2 × PT6T turboshaft   *Rotor dia:* 14.69m   *Length:* 12.92m

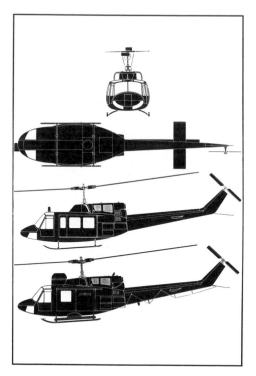

The Model 212 Twin Two-Twelve 15-seat military/civil utility helicopter first flew in 1968. Military aircraft are designated UH-1N in the USA and CH-135 in Canada. Powerplant is a 1,290shp Pratt & Whitney of Canada PT6T-3 Turbo Twin-Pac. The Agusta-Bell Model 212 is produced in Italy by Agusta, the 212ASW being for the Italian Navy, Turkey, Iran and a number of other countries. Main external difference is the large cylindrical radome mounted on top of the forward cabin roof. Armament options include two homing torpedoes, or depth charges, or wire guided missiles. The US 212 line was transferred to Canada in 1988. A four blade rotor development of the 212 is the 412 of which the latest version is the 412SP. *Countries of origin:* USA/Italy/Canada. *Main silhouette:* Bell 212; *lower side view:* Agusta 212ASW. *Picture:* 412SP.

Bell Model 205/UH-1 Iroquois/Model 205A-1

Power: 1 × T53 turboshaft   *Rotor dia:* 14.63m   *Length:* 12.77m

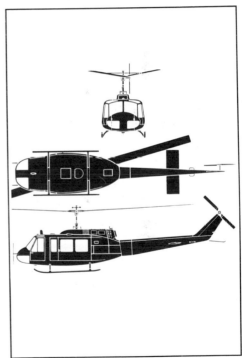

The Model 205 military general-purpose helicopter first flew in 1961. Basically similar to the earlier Model 204 Iroquois, it has a longer fuselage, an additional cabin window and equipment improvements. The Model 205A-1 15-seat commercial utility version of the Model 205 is powered by a 1,250shp Lycoming turboshaft. Both versions have also been built under licence in Italy by Agusta. Licence production is also undertaken in Japan (HU-1H) Taiwan and Turkey. Cruising speed for all versions is about 200km/h. *Country of origin:* USA. *Main silhouette:* UH-1D. *Picture:* UH-1H.

Bell Model 214A/214B/214ST *Confusion:* Bell 205

Power: 1 × T55 or 2 × CT7 turboshaft   *Rotor dia:* 15.24m   *Length:* 12.92m

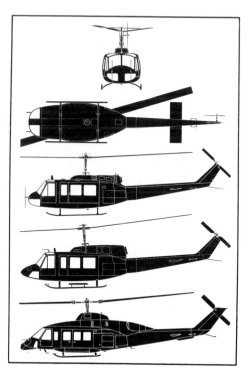

Initially produced in quantity for Iran as the Isfahan, the Bell 214A military utility helicopter was developed from the Bell 205 and first flew in 1974. Iran also ordered the 214C version for search and rescue. The civil version is the 214B BigLifter, which carries 14 passengers or can be used for cropspraying or firefighting. The latter version is known as the BigFighter. Typical 214 internal load is 1,814kg, while up to 3,175kg can be slung externally. The BigLifter differs from the 214A in having a fire-extinguishing system, push-out escape windows and commercial avionics. Agusta has also built the 214A. Cruising speed is 259km/h and range 483km. Also produced in quantity is the 214ST Supertransport with two GE CT7 engines, longer fuselage and accommodation for up to 18 passengers. Over 80 Super Transports have been built. *Country of origin:* USA. *Main silhouette:* Model 214; *middle side view:* Model 214B; *lower side view:* Model 214ST. *Picture:* Model 214ST.

Bell Model 206 JetRanger/OH-58 Kiowa

Power: 1 × Allison 250 turboshaft   *Rotor dia:* 10.16m   *Length:* 9.5m

This general-purpose and light observation helicopter first flew in 1962. The original Model 206A JetRanger remained in production until 1972, when 660 had been built. Deliveries of the Model 206B JetRanger II, powered by a 400shp Allison turboshaft, began in 1971. A total of 1,619 were delivered. Current production model is the Model 206B JetRanger III, which has an uprated engine of 420shp, and enlarged tail rotor mast. The Kiowa is the military variant of the JetRanger, from which it differs in having a main rotor of slightly greater diameter and equipment changes for its military role. Bell is converting US Army Kiowas to OH-58D standard with mast-mounted sight. An armed version is known as the Warrior. A new development is the light, simplified combat helicopter, the Model 406 Combat Scout. *Country of origin:* USA. *Silhouette:* OH-58A Kiowa. *Picture:* 406 Combat Scout.

MBB-Kawasaki BK 117 *Confusion:* JetRanger

Power: 2 × LTS101 turboshafts   *Rotor dia:* 11m   *Length:* 13m

A joint development by West Germany and Japan, the BK 117 seats seven passengers in executive form and up to eleven in high-density transport form. With a range of 545km, the BK 117 has a cruising speed of 234km/h. The first prototype flew in Germany in May 1979, followed by a second in Japan. There are two production centres, at Gifu in Japan and Munich in Germany. The BK117 is also being built in Indonesia. Latest production version is BK 117B. Over 170 BK 117s have been delivered. *Country of origin:* Japan/West Germany.

Power: 2 × **Gem turboshafts**   *Rotor dia:* **12.8m**   *Length:* **12.06m**

One of the three helicopters covered by the Anglo-French agreement of 1967, the Lynx first flew in 1971. Two main versions have been built, one for naval use and the second for the British Army. Equipped with a skid undercarriage, the Lynx AH.1 entered British Army service in 1977. Other British Army versions are AH.5/7/9. Naval versions are the Lynx HAS.2, HAS.3 and HAS.8 for the Royal Navy, and variants for the French Navy, the Royal Netherlands Navy and the navies of Argentina, West Germany and Brazil. The type has also been ordered by Qatar, Denmark and Norway. All naval aircraft and the new Lynx AH-9 are equipped with a wheeled undercarriage. The latest version is the naval Super Lynx with higher powered engines and modified avionics, main and tail rotor. *Countries of origin:* UK/France. *Main silhouette:* AH.1; *lower side view:* HAS2. *Picture:* **Super Lynx**.

Bell Model 206L LongRanger

Confusion: Bell 206, Lynx

Power: 1 × Allison turboshaft   *Rotor dia:* 11.28m   *Length:* 10.13m

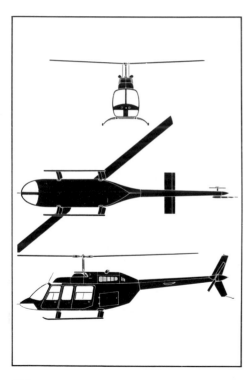

The Model 206L LongRanger seven-seat general-purpose helicopter first flew in 1974. Developed from the Model 206B JetRanger II, it has a longer fuselage with an additional pair of cabin windows. First deliveries were in 1975. Current production version is the LongRanger II. Double doors on the port side of the cabin provide an opening 1.52m in width, permitting the straight-in loading of stretchers or cargo. Two stretchers and two walking patients can be carried in the ambulance role, while an optional executive interior accommodates four passengers. Powered by a 420shp Allison turboshaft, the LongRanger cruises at 210km/h over a sea-level range of 550km. Current production version, in Canada, is the Model 206L-3 LongRanger III. *Country of origin:* USA. *Silhouette:* Model 206L. *Picture:* LongRanger III.

Power: 2 × Allison 250 turboshafts or 2 Arriel turboshafts   *Rotor dia:* 11m   *Length:* 13.05m

The Agusta A109A high-speed twin-engined helicopter is available in both civil and military versions. First flown in 1971, the type entered production in 1976. The A109 is powered by two 420shp Allison turboshafts and features a fully retractable undercarriage. Military versions, under construction for the armies and navies of several nations, are externally similar to the civil model, varying only in the armament and equipment carried. An improved Mk II version is now on the line. The military Mk II can perform the utility, ECM, ambulance, scout/attack/air defence and anti-tank roles. The A109 EOA is for observation work. A new military variant is the A109K with two Arriel engines, fixed undercarriage and longer nose. *Country of origin:* Italy. *Silhouette:* A109A. *Picture:* A109K.

Power: 2 × LTS101 turboshafts   *Rotor dia:* 11.89m   *Length:* 10.98m

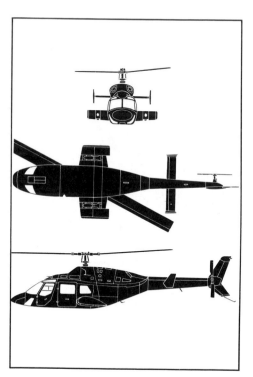

The Model 222 eight-seat light commercial helicopter first flew in 1976. The first commercial light twin-engined helicopter to be built in the USA, the Model 222 has a fully retractable undercarriage accommodated in two short-span sponsons set low on each side of the fuselage. The tailplane, mounted midway along the rear fuselage, is fitted with endplate fins. The type is produced in three versions: the basic Model 222B, the 222B Executive and the 222UT. Externally alike, these variants are distinguished by equipment differences except on the 222UT which has a skid undercarriage. Powered by two 600shp Avco Lycoming turboshafts, the Model 222 cruises at 265km/h. *Country of origin:* USA. *Picture:* Model 222UT.

Power: 2 × T58 turboshafts   *Rotor dia:* 13.41m   *Length:* 11.68m

The H-2 Seasprite anti-submarine, anti-missile defence, search and rescue, observation and utility helicopter first flew in 1959. The original series was single-engined and, as the UH-2A, entered service with the US Navy in 1963. Current versions are twin engined and include the HH-2D for coastal and geodetic survey; NHH-2D test aircraft. SH-2D LAMPS anti-submarine and utility aircraft; and SH-2F, a developed LAMPS version. All SH-2Ds are being upgraded to SH-2F standard. Latest version is the SH-2G Super Seasprite with T700 engines and improved avionics. *Country of origin:* USA. *Silhouette:* SH-2F. *Picture:* SH-2D.

Westland 30

Power: 2 × Gem turboshafts   *Rotor dia:* 13.31m   *Length:* 14.33m

A combination of the Lynx rotor and powerplant with a new transport fuselage, the Westland 30 seats 17 passengers to airliner standards and up to 22 in high-density layout. The type is also available as a multi-role military helicopter. First flown in April 1979, the Westland 30 ceased production in 1988 after 38 had been built. The initial production version was the 30 Series 100, the 100-60 has Gem 60 engines and the Series 200 has GE CT7 engines. Cruising speed is 241km/h and range with seven passengers is 686km. *Country of origin:* UK.

Power: 1 × Allison 250 turboshaft   *Rotor dia:* 8.03m   *Length:* 7.01m

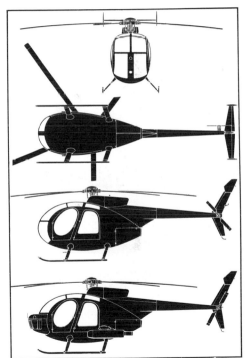

The Model 500 series is the counterpart of the US Army's OH-6 Cayuse. The 500 was originally a Hughes product. The Model 500 and Model 500C are identical apart from the installation of the uprated 400shp Allison turboshaft in the Model 500C. The Model 500D and Model M-D are distinguished by a small T-tail and five-blade rotor. The Model 500D is the basic commercial version, while the 500M-D Defender is the multi-role military variant. The 500E has a longer and more streamlined nose while the 530F Lifter is for high temperature/high altitude operations. Latest model is the 530MG with mast-mounted sight. The Nightfox is for night surveillance and the MH-6A for special operations. *Country of origin:* USA. *Main silhouette:* Model 500D; *lower side view:* 500MD Defender. *Picture:* 530MG.

Power: 2 × Allison 250 turboshafts   *Rotor dia:* 9.84m   *Length:* 11.86m

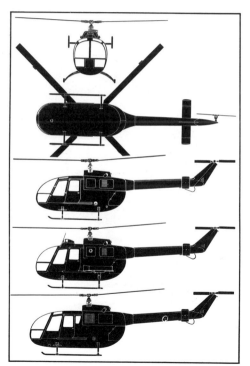

The MBB BO 105 five-seat light utility helicopter first flew in 1967. A number of different versions have been built, but all are essentially similar in external appearance. Original standard production model was the BO 105C, powered by two 400shp Allison turboshaft engines. These were succeeded in 1975 by the BO 105CB, powered by two 420shp Allison turboshaft engines and incorporating strengthened rotor gearing. Other versions currently in production are the BO 105CBS with longer fuselage and increased seating capacity; BO 105D for the UK market; BO 105M (PAH 1) military version armed with six HOT missiles; the BO 105M (VBH) light observation helicopter and the 105P (PAH-1) anti-tank variant. BO105LS is a version produced in Canada. *Country of origin:* West Germany. *Main silhouette:* BO 105CB; *middle side view:* BO 105P; *bottom side view:* BO 105CBS. *Picture:* BO 105C.

Aérospatiale AS.350/355 Ecureuil/AStar/TwinStar

Power: 1 × Arriel or 1 × LTS101   *Rotor dia:* 10.69m   *Length:* 10.91m
or 2 × Allison 250 turboshafts

Designed as a successor to the Alouette II, the Ecureuil (Squirrel) first flew in 1975 as the Arriel-powered AS350B. The Lycoming-powered AS.350C AStar was intended for the US market alone, with the AS350B available elsewhere. Accommodation is provided for six people in bucket seats in the front of the cabin and two rows of bench seats behind. Performance of the two versions is similar, cruising speed being 235km/h. A twin-engined version is known as the TwinStar (AS355E) in the USA and the Ecureuil 2 (AS355F) elsewhere. The 355N has two TM319 engines. A further military version is the A5355M. *Country of origin:* France. *Silhouette:* AS 350B Ecureuil. *Picture:* AS355M.

Rogerson RH-1100 *Confusion:* AS 350B

Power: 1 × T63 turboshaft   *Rotor dia:* 10.79m   *Length:* 9.08m

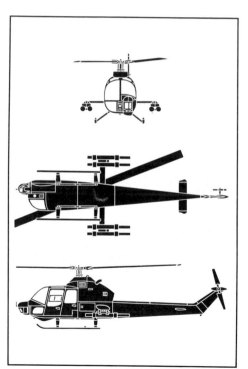

Originally known as the Hiller FH-1100, this utility helicopter was the first American light turbine-powered helicopter to enter the market, evolved from the Hiller OH-5A, which first flew in 1963. The FH-1100 was developed by Fairchild Industries, which had taken over Hiller and has in turn been taken over by Rogerson Aircraft Corp. Production eventually reached about 250, for both civil and military operators, and the type remains in small-scale service with the armed forces of a number of South American nations. Versions currently on offer by Rogerson are the armed RH-1100M Hornet and the RH-1100. *Country of origin:* USA. *Silhouette:* RH-1100M Hornet. *Picture:* RH-1100M Hornet.

Power: 1 × Lycoming piston engine   *Rotor dia:* 7.24m   *Length:* 6.62m

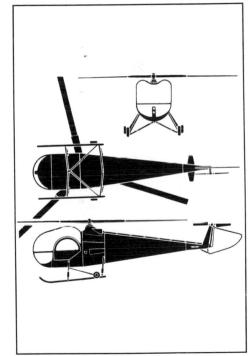

The Model B-2 two-seat light helicopter was first flown in 1953, though deliveries did not begin until 1959. The B-2A is an updated version with redesigned cabin and improved equipment. Main production version is the B-2B, which differs in having a fuel-injected engine for improved performance. Powered by a 180hp Lycoming piston engine, the B-2B has a cruising speed of 145km/h. Hynes Helicopter Inc put the H-12 (formerly B-2B) back into production as the Hynes H-2; and subsequently sold production rights to Naras Aviation in India. *Country of origin:* USA. *Silhouette:* B-2B. *Picture:* B-2.

 **Brantly-Hynes 305**   *Confusion:* Brantly-Hynes B-2

Power: 1 × Lycoming piston engine   *Rotor dia:* 8.74m   *Length:* 7.44m

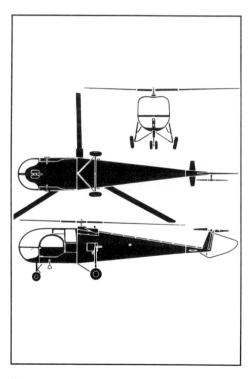

The Model 305 five-seat light helicopter first flew in 1964. Similar in configuration to the Model B-2B, the Model 305 is larger in all respects. Accommodation consists of two front seats side-by-side, with a rear bench seat for three persons. The Model 305 can be fitted with skid, wheel or float undercarriage. Powered by a 305hp Lycoming flat-six piston engine, the Model 305 cruises at about 175km/h. Manufacturing rights for the 305 have now been acquired by Naras which intends to build the type at a factory near Madras, India. *Country of origin:* USA.

Power: 1 × Lycoming piston engine   *Rotor dia:* 9.75m   *Length:* 8.94m

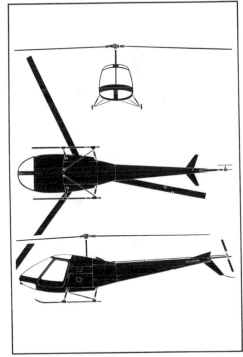

The original three-seat F-28 first flew in 1962 and entered production in 1965. During 1973 an advanced version of the basic Model F-28A was developed. Known as the Model 280 Shark, it is generally similar to the F-28A but has improved cabin contours and some small modifications to the tail area. There are a number of production versions: the F-28A powered by a 205hp Lycoming; Model 280 Shark; F-28C with turbocharged 205hp Lycoming; Model 280C; F-28C-2; and Models 280C, 280F and 280FX Hawk, 28F-P Sentinel and F28F Falcon. *Country of origin:* USA. *Silhouette and picture:* Model 280.

Silvercraft SH-4 *Confusion:* Shark

Power: 1 × Franklin piston engine   *Rotor dia:* 9.03m   *Length:* 7.65m

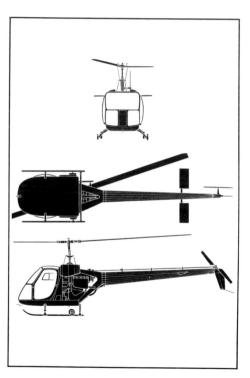

The SH-4 is a three-seat light helicopter designed for pilot training, utility, agricultural and similar roles. First flown in 1965, the type entered full-scale production in 1967 and a total of 50 were completed. Two versions were built: the standard SH-4 and the SH-4A agricultural version fitted with spraybars and two chemical tanks. Powered by a 235hp Franklin engine derated to 170hp, the SH-4 cruises at 130km/h. Other performance figures include a maximum level speed at sea level of 161km/h and a range of 320km. *Country of origin:* Italy. *Silhouette:* SH-4. *Picture:* SH-4A.

Power: 1 × T53 or 1 × T400 turboshaft   *Rotor dia:* 13.41m   *Length:* 13.59m

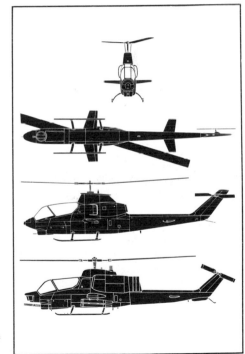

The Model 209 HueyCobra armed attack helicopter first flew in 1965. Four single-engined versions have been built to date: AH-1G, the original version for the US Army; AH-1Q anti-armour version with eight Tow missiles; AH-1R, as AH-1G but with a more powerful engine; and the AH-1S advanced version equipped with Tow missiles and featuring a flat-plate canopy. Some 690 AH-1Gs will ultimately be modified to AH-1S standard. AH-1S has a cruising speed of 225km/h. The twin-engined versions of the HueyCobra are the AH-1J SeaCobra, operated by the US Marine Corps, and the AH-1T Improved SeaCobra. Latest version is the AH-1W SuperCobra for use by the US Marine Corps. *Country of origin:* USA. *Main silhouette:* AH-1G; *lower side view:* AH-1W. *Picture:* AH-1S.

Agusta A129 Mangusta

Confusion: Bell AH-1 HueyCobra

Power: 2 × Gem turboshafts   *Rotor dia:* 11.9m   *Length:* 14.29m

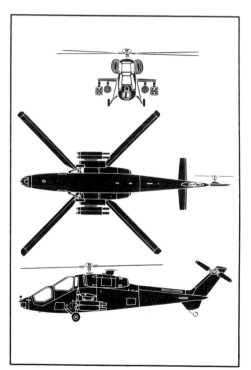

Deliveries of the A129 Mangusta light anti-armour helicopter to the Italian Army began in 1988. Powered by two Rolls-Royce Gem engines, the A129 is a two-seater with a maximum speed of 259km/h and a maximum endurance of 3 hours. Armament is carried on four underwing mountings and can consist of eight TOW or six Hellfire anti-tank missiles. Other weapons include HOT missiles, two gun pods or rocket projectiles. The A129 first flew in September 1983. A light battlefield (LBH) version of the A129 is proposed. Agusta, Westland, Fokker and CASA are working on an A129 development known as Tonal. *Country of origin:* Italy.

Power: 1 × Lycoming piston engine   *Rotor dia.* 7.71m   *Length:* 6.8m

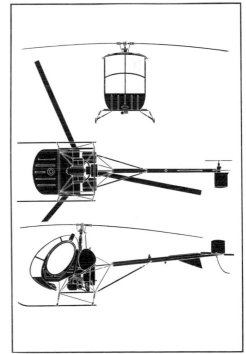

Design and development of the original Hughes Model 269 two-seat light helicopter began in 1955, with the first aircraft flying the following year. Production aircraft were designated Model 269A and by early 1968 nearly 1,200 had been built for both civil and military operators. The Model 300 was developed under the designation Model 269B, with production commencing in 1964. In 1964 the Model 269A was selected by the US Army as a light helicopter primary trainer and designated TH-55A Osage. Production was transferred to Schweizer in 1983 and deliveries of the 300C continue. *Country of origin:* USA. *Silhouette:* Model 300. *Picture:* TH-55A.

Mil Mi-2 'Hoplite'

Confusion: —

Power: 2 × Isotov turboshafts   *Rotor dia:* 14.5m   *Length:* 17.42m

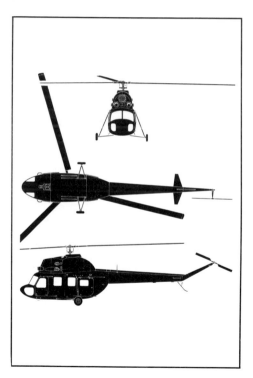

The Mi-2 twin-engined development of the Mi-1 was first announced in late 1961. Under an agreement signed in 1964, production and marketing were assigned to WSK-PZL Swidnik in Poland. Five thousand have since been built, mainly for export, with more than 2,000 being delivered to the USSR. The type is used by both civil and military operators and can be fitted out for a variety of roles, including ambulance, search and rescue, agricultural work, freight and transport of up to eight passengers. The agricultural version is called the Bazant. Mi-2s of the Polish Air Force have been fitted with unguided rocket pods and air-to-surface missiles on pylons on each side of the fuselage. *Country of origin:* USSR/Poland.

Aérospatiale AS 332L Super Puma

Boeing Vertol CH-47 Chinook

Kamov Ka-25 'Hormone B'

Mil Mi-4 'Hound'

Confusion: S55

Power: 1 × Shvetsov piston engine   *Rotor dia:* 21m   *Length:* 16.8m

The Mi-4, code-named 'Hound', was first put into production in 1952 and several thousand examples have been built for military and civil work. Exports have been widespread throughout the Soviet sphere of influence and to countries such as Egypt and India. Fourteen troops can be carried in the military version, while freight can be loaded through the rear clamshell doors. A military close-support variant is armed with a machine gun in an underfuselage pod, plus air-to-surface rockets. The anti-submarine version is fitted with magnetic anomaly detection gear and an undernose search radar. The civil version is the 11-passenger Mi-4P and the agricultural variant is the Mi-4S. China built the type under licence. Economical cruising speed is 160km/h and maximum range is 595km. *Country of origin:* USSR. *Silhouette:* Mi-4P. *Picture:* Mi-4 (military).

Power: 1 × R-1300 piston engine   *Rotor dia:* 16.15m   *Length:* 12.88m

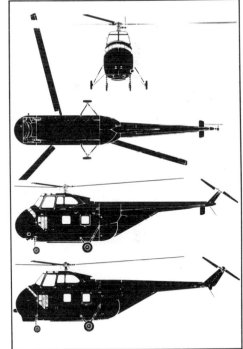

The S-55, the first American helicopter to be certificated for commercial operation, first flew in 1949. The type has been sold to civil and military operators all over the world and has also been licence-built in Japan by Mitsubishi and in Britain by Westland as the Whirlwind. The military version is designated H-19 Chickasaw in US service use and is still operated by Chile and Turkey. More than 1,200 were built. Powered by an 800hp Wright R-1300 piston engine, the S-55 has a cruising speed of about 145km/h. US company Helitec has produced a turbine-powered version, the S-55T. *Country of origin:* USA. *Main silhouette:* S-55; *lower side view:* S-55T. *Picture:* S-55T.

 **Sikorsky S-58/S-58T** *Confusion:* S-55, Wessex

Power: 1 × R-1820 piston engine or
1 × PT6T turboshaft

Rotor dia (S-58T): 18.9m   *Length:* (S-58T): 16.69m

A general-purpose 18-passenger piston-engined military/civil helicopter, the S-58 has been operated all over the world and production totalled 1,821. US Army designations are CH-34A to CH-34D Choctaw. The marine version is known as the UH-34D, VH-34D and UH-34E Seahorse, and the name Seabat is used for anti-submarine variants (SH-34G and SH-34J). The S-58B, C and D are civil transports. First flown in 1970 with two PT6 turboshaft engines, the S-58T was eventually offered as new or in the form of a conversion from the piston-engined version. *Country of origin:* USA. *Main silhouette:* S-58; *lower side view:* S-58T. *Picture:* S-58T.

Power: 2 × Gnome turboshafts   *Rotor dia:* 17.07m   *Length:* 14.74m

A general-purpose and anti-submarine helicopter first flown in 1958, the Wessex was originally powered by a Gazelle engine but was later fitted with Rolls-Royce Gnomes. Used by the RAF, Royal Navy and Royal Australian Navy, the Wessex has been built in various versions, including the HC2 transport/ambulance for the RAF and the HAS1 anti-submarine and HU5 commando assault transport for the Royal Navy. The Wessex carries a crew of two and ten passengers. Range is 538km. *Country of origin:* UK. *Silhouette:* HAS1. *Picture:* Wessex HC2.

Aérospatiale/Westland SA.330 Puma *Confusion:* Wessex, Super Frelon, Sea King

Power: 2 × Turmo turboshafts   *Rotor dia:* 15m   *Length:* 14m

Originally a French product which first flew in 1965, the SA.330 Puma was later adopted as part of a joint Franco-British programme under which production was shared between the two countries. The RAF and the French Army use the type as a transport and some 700 have been sold for military and civil applications in many countries. Early versions were the 330B (French Army), 330C (export), 330E (RAF), 330F and G (civil) and 330H (military). Current models are the 330J (civil) and 330L (military). Sixteen troops or up to 20 passengers can be carried at a cruising speed of 257km/h. The Puma is licence-built in Romania. *Countries of origin:* France/UK/Romania. *Silhouette:* SA.330B. *Picture:* Puma HC1 (SA.330E).

Aérospatiale SA.332 Super Puma

Power: 2 × Makila turboshafts   *Rotor dia.* 15.08m   *Length:* 14.82m

A much modified version of the Puma, the 332 Super Puma has a lengthened nose, new undercarriage and higher-powered engines. The five versions of the 332 on offer are the military 332B seating 20 troops; civil 332C seating 17; 332F for ASW/search and rescue and other naval duties; 332L with 0.76m fuselage extension, four more seats and two extra windows; and 332M with the same extension for military use. More than 275 Super Pumas had been ordered by spring 1988. Thirty-five 332Ls for Bristow Helicopters are known as Tigers. The 332L has a maximum speed of 296km/h and range at cruising speed of 848km. *Country of origin:* France. *Picture:* 332 M1.

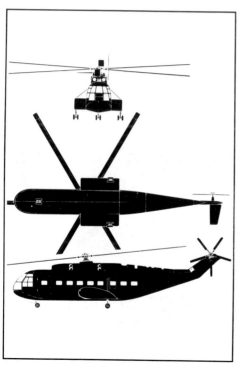

Aérospatiale SA.321 Super Frelon

Confusion: Puma, Super Puma, S-61L/N

Power: 3 × Turmo turboshafts   *Rotor dia:* 18.9m   *Length:* 23m

First flown in 1962, the three-engined SA.321 Super Frelon is used as a military and civil transport and for anti-submarine work. Civil versions are the 321F and 321J, the former designed to carry 34/37 passengers. The 321G for the French Navy is amphibious, has radar and carries torpedoes and other military equipment. The 321H army/air force variant accommodates 27–30 troops or cargo. The Super Frelon has been built in some numbers and sold to eight countries, including Israel. Cruising speed is 250km/h and range 820km. *Country of origin:* France. *Silhouette:* SA.321F. *Picture:* SA.321K.

Power: 2 × CT58 turboshafts   *Rotor dia:* 18.9m   *Length:* 22.2m

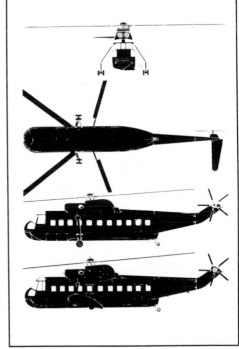

A widely used helicopter airliner, the S-61L/N can seat up to 28 passengers. The S-61L is non-amphibious, while the N has stabilising floats and a sealed hull for use on water. British International Helicopters is the largest single operator of this type. The S-61L first flew in 1960 and the S-61N in 1962. The undercarriage is retractable and the nose carries a thimble radome containing weather radar. S-61N production rights are now held by Agusta. This company has now produced the AS-61N1 Silver with 1.27m longer fuselage, re-arranged windows and smaller sponsons. Range is 796km and cruising speed 222km/h. *Country of origin:* USA. *Main silhouette:* S-61L; *lower side view:* S-61N. *Picture:* Agusta AS-61N1 Silver.

Sikorsky S-62

Confusion: S-61L/N

Power: 1 × T58 turboshaft   *Rotor dia:* 16.16m   *Length:* 13.58m

The first amphibious helicopter to be built by Sikorsky, the S-62 first flew in 1958 and has been used in both military and civil forms. As an airliner it carries 12 passengers or cargo. The original S-62A was licence-built by Mitsubishi in Japan, the S-62B had the main rotor system of the S-58 instead of that of the S-55, and the S-62C was the last commercial version. The US Coast Guard took delivery of 99 S-62s, designating them HH-52A. Maximum cruising speed is 158km/h and range 764km. *Country of origin:* USA. *Silhouette and picture:* HH-52A.

Power: 2 × T58 turboshafts   *Rotor dia:* 18.9m   *Length:* 16.69m

The multi-purpose S-61 is used for anti-submarine work, transport and search and rescue. As the S-61 the type can carry 26 troops. In the anti-submarine role it is known as the Sea King and carries weapons, including homing torpedoes. US Navy designation is SH-3 and there are a number of variants, from SH-3A through to SH-3H. The SH-3 is known as the CH-124 in the Canadian Armed Forces. The SH-3 has been widely exported and is standard equipment in Japan and Italy, where it is built by Agusta. The SH-3 first flew in 1959. Range is 1,005km and cruising speed 219km/h. *Country of origin:* USA. *Silhouette:* SH-3H. *Picture:* SH-3D.

Power: 2 × Gnome turboshafts   *Rotor dia:* 18.9m   *Length:* 17.01m

Developed under licence from the American S-61D, the Sea King is the Royal Navy's standard anti-submarine helicopter under the designations HAS1 and 2. The HAR3 is used by the RAF for search and rescue. The Sea King has been exported widely. The Commando military transport version can seat 30 troops and has no amphibious capability. Commando Mks 1, 2, 3 and HC4 are in service. The Commando Mk 1 differs minimally from the Sea King, while the Mk 2, the main production version, can be used for tactical troop transport, logistic support, casualty evacuation, air-to-surface strike, and search and rescue. The Mk 42 is for India, the HAS Mk 5 for the Royal Navy and several have been fitted with Searchwater radar as the Sea King AEW. Sea King HAS Mk6 is the latest updated version for the Royal Navy. *Country of origin:* UK. *Silhouette:* HAS1. *Picture:* Sea King AEW2.

Sikorsky S-61R/CH-3E/HH-3F Pelican

Power: 2 × T58 turboshafts   *Rotor dia:* 18.9m   *Length:* 17.45m

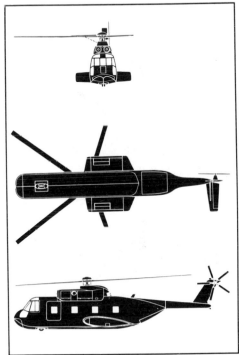

Developed from the S-61 Sea King, the S-61R has a rear loading ramp and retractable undercarriage. It is used for transport, assault and search and rescue. The first S-61R flew in 1963 and the type is used by the USAF as the CH-3E and HH-3E and by the US Coast Guard as the HH-3F Pelican, with a nose radome. The HH-3E has armour, defensive armament and an in-flight refuelling probe on the starboard side. Up to 30 troops can be carried and a hoist is fitted. Agusta in Italy continues to build a version of the Pelican for the Italian Air Force. Cruising speed is 232km/h and range 748km. *Country of origin:* USA. *Silhouette:* HH-3F. *Picture:* HH-3E.

 **EHI EH 101** *Confusion:* S-61R, S-65A

Power: 3 × T700 OE RTM 322 turboshaft engines   *Rotor dia:* 18.59m   *Length:* 22.81m

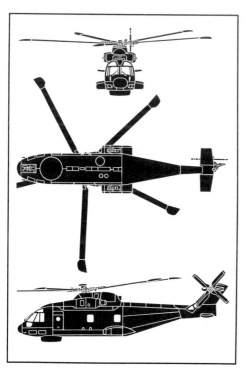

Intended for naval, military and commercial use, the EH 101 represents a combined design and development effort by Westland in the United Kingdom and Agusta in Italy. First flown in October 1987, the EH 101 is to be used in the maritime role by the British, Italian and Canadian navies. In commercial form the EH 101 will seat 30 passengers, while as a military tactical transport it will carry 35 fully equipped troops. Basic engine is the General Electric T 700 while, at the time of writing, the Rolls Royce RTM322 was being considered as an alternative. In its naval form the EH 101 will carry advanced avionics and anti-submarine systems and will serve aboard vessels from frigate size upwards. The naval variant has a maximum take-off weight of 13,000kg and a cruising speed of 259km/h. *Countries of origin:* UK/Italy.

Power: 2 × Isotov turboshafts   *Rotor dia:* 21.29m   *Length:* 18.3m

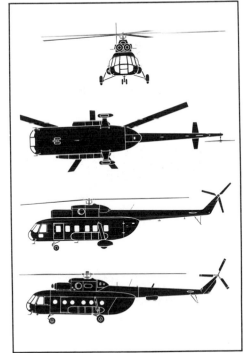

A replacement for the Mi-4 'Hound', the Mi-8 is a twin-turbine helicopter seating up to 32 passengers. As with other Soviet medium helicopters, the 'Hip' is used by both the Soviet services and Aeroflot. The military versions have round windows and are code-named 'Hip C' to 'Hip K', this last version being used for ECM work. 'Hip E' carries a gun and massive rocket-projectile armament. The transport has rectangular windows and no radome. With up-rated engines and other modifications a variant of the Mi-8 is known as the M-17, code-named 'Hip H'. Range is 375km and maximum cruising speed 225km/h. Over 10,000 'Hips' have been built. *Country of origin:* USSR. *Main silhouette:* 'Hip C'; *lower side view:* 'Hip H'. *Picture:* 'Hip H'.

 **Mil Mi-14 'Haze'** *Confusion:* 'Hip', S-61

Power: 2 × Isotov turboshafts *Rotor dia:* 21.29m *Length of fuselage:* 18.15m

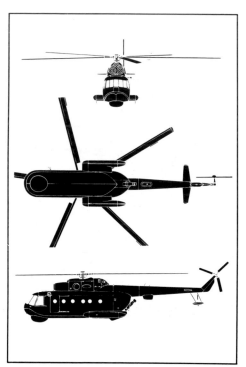

Developed from the Mi-8, the Mi-14 'Haze' is a shore-based Soviet Navy anti-submarine helicopter. A boat hull is incorporated, there is a radome under the nose, and a magnetic anomaly detector is mounted under the tailboom. The undercarriage is fully retractable. 'Haze' is in full production and in large-scale service with the Soviet Naval Air Force. Over 100 have been delivered. Dimensions and dynamic components are generally similar to those of the Mi-8, while the powerplant is a pair of Isotov TV3-117 turboshafts of the kind fitted to the Mi-24 gunship. There are three versions, Mi-14PL 'Haze A' for ASW, Mi-14BT 'Haze B' for mine countermeasures and Mi-14PS 'Haze C' for search and rescue. Maximum speed is approximately 230km/h. *Country of origin:* USSR. *Silhouette:* 'Haze A'. *Picture:* 'Haze A'.

Power: 2 × Isotov turboshafts   *Rotor dia:* 16.76m   *Length:* 16.9m

A heavily armed assault and gunship helicopter, the Mi-24, code-named 'Hind', is in service in the Soviet Union in very large numbers. 'Hind A' carries eight troops and has stub-wing-mounted missiles and a nose gun, while 'Hind D' has separate cockpits for pilot and weapon operator and a four-barrel machine gun in a chin mounting. 'Hind C' has no nose gun or under-nose sighting system. 'Hind E' carries tube-launched anti-tank missiles. 'Hind F' has twin gun side pack and no nose turret, and 'Hind G' is used for radiation sampling. Early 'Hinds' had the tail rotor on the starboard side, while later versions have it on the port side. *Country of origin:* USSR. *Silhouette:* 'Hind D'. *Picture:* 'Hind E'.

Power: 2 × Allison 250 or PT6B turboshafts   *Rotor dia:* 13.41m   *Length:* 13.44m

A commercial transport seating 14 passengers, the Sikorsky S-76 first flew in 1977. The S-76 is classified as an all-weather helicopter. Maximum cruising speed at a gross weight of 3,810kg is 286km/h. Range with eight passengers, auxiliary fuel and offshore equipment is 1,112km. Current production version is the S-76 Mk II. In addition there are the S-76 Utility, the S-76B with PT6B engines, the AUH-76 armed utility model and the H-76N naval variant. *Country of origin:* USA. *Silhouette:* S-76A. *Picture:* S-76B.

McDonnell Douglas AH-64A Apache

Power: 2 × T700 turbofans   *Rotor dia:* 14.63m   *Length:* 14.68m

By mid 1988 360 AH-64A Apache tandem two-seat all-weather attack helicopters had been delivered. Powered by two 1696shp GE T700 engines, the Apache is capable of a maximum speed of 300km/h. The helicopter mounts a 30mm Chain Gun automatic cannon under the fuselage and can carry rockets or 16 Hellfire missiles. An advanced sighting and fire control system is fitted. The first prototype flew in September 1975. The type was originally called Hughes AH-64A, prior to the company's take-over by McDonnell Douglas. *Country of origin:* USA.

559

Sikorsky S-70/UH-60A Black Hawk   *Confusion:* S-76, Seahawk

Power: 2 × T700 turboshafts   *Rotor dia:* 16.23m   *Length:* 15.26m

First flown in 1974, the Sikorsky S-70 Black Hawk, military designation UH-60A, was the winner of the US Army's Utility Tactical Transport Aircraft System (Uttas) competition and is in full-scale production. Ultimately over 1,000 Black Hawks will be acquired. Eleven fully equipped troops can be carried, and alternative loads include cargo and stretchers. Up to 3,629kg can be carried externally on a hook. The HH-60A Night Hawk is a combat rescue version for the US Air Force, while the EH-60A is equipped for electronic countermeasures. S-70C is a commercial variant. Westland is building the S-70 in the UK as the WS70. *Country of origin:* USA/UK. *Silhouette:* UH-60A. *Picture:* WS70.

Power: 2 × T700 turboshafts   *Rotor dia:* 16.36m   *Length:* 15.26m

Developed from the Black Hawk, the three-crew SH-60B Seahawk was the winner in the US Navy's Light Airborne Multi-Purpose System (LAMPS) Mk III competition. The Seahawk is used for anti-submarine warfare and ship surveillance/ship targeting from cruisers, destroyers and frigates. Sophisticated detection and processing equipment and anti-submarine torpedoes are carried. Deliveries started in 1983. The Seahawk has also been purchased by Japan and Australia. Other versions are SH-60F Ocean Hawk, HH-60 and HH-60J. *Country of origin:* USA.

Sikorsky S-65A/CH-53 Sea Stallion *Confusion:* CH-53E

Power: 2 × T64 turboshafts   *Rotor dia:* 22.02m   *Length:* 20.47m

A heavy assault transport, the S-65A first flew in 1964 and incorporates many S-64 components. Fitted with rear loading doors, the type can carry vehicles, guns or 55 troops. The US Marines and Iran use the CH-53A Sea Stallion, while the USAF has the HH-53B/C. An improved US Marine Corps version is the CH-53D, and the CH-53G is licence-built in Germany. The S-65 (MCM), designated RH-53D in the US Navy, is used for mine countermeasures and has a tow boom. The helicopter will remain in USMC service until at least 1995. Range is 410km and maximum cruising speed 278km/h. *Country of origin:* USA. *Silhouette:* CH-53G. *Picture:* CH-53D.

Sikorsky CH-53E Super Stallion

Power: 3 × T64 turboshafts   *Rotor dia:* 24m   *Length:* 22.38m

Developed from the S-65/CH-53A/B/D Sea Stallion, the CH-53E Super Stallion three-engined amphibious assault transport helicopter is used by both the US Navy and US Marines. Up to 56 troops can be accommodated, and very large loads, such as aircraft, can be slung externally. The first prototype flew in 1974. An airborne mine countermeasures version is the MH-53E SeaDragon, which has been delivered to US Navy. These have larger sponsons and hydrofoil magnetic sweeping gear. Maximum cruising speed is 278km/h. *Country of origin:* USA. *Silhouette:* CH-53E. *Picture:* CH-53E.

 **Mil Mi-6 'Hook'** *Confusion:* 'Harke'

Power: 2 × Soloviev turboshafts   *Rotor dia:* 35m   *Length:* 33.18m

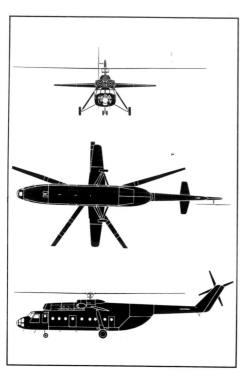

When it first appeared in 1957 the Mi-6, code-named 'Hook', was the world's largest helicopter. Basically a military design, of which over 800 have been built, the type has also been employed by Aeroflot as a specialised civil transport. Military 'Hooks' have been exported to Egypt, Bulgaria and Vietnam. Equipment, including missiles, can be loaded through the rear clamshell doors and heavy loads can be carried externally. A total of 65 passengers can be accommodated. Maximum cruising speed is 250km/h and range 650km. At one time the Mi-6 held a total of 14 FAI-recognised records. The Mi-6 is in service in eight countries. *Country of origin:* USSR.

Power: 2 × Lotarev turboshafts   *Rotor dia:* 32m   *Length:* 33.7m

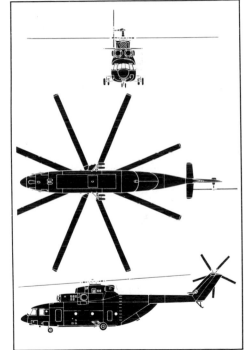

The heaviest helicopter built anywhere in the world, the Mi-26, NATO code-named 'Halo', is the first to have an eight-blade rotor. Initially flown in 1977 and now in full-scale production, the Mi-26 can carry 20,000kg of internal payload or up to 85 troops. Capable of operating in extreme winter conditions in the Soviet Union in the civil role, the Mi-26 also has a variety of military applications. The Mi-26 made its first public appearance at the 1981 Paris Show, where it aroused major interest. The Mi-26's top speed is 295km/h and range 800km. *Country of origin:* USSR.

Mil Mi-10/Mi-10K 'Harke' *Confusion:* 'Hook', Skycrane

Power: 2 × Soloviev turboshafts   *Rotor dia:* 35m   *Length:* 32.86m

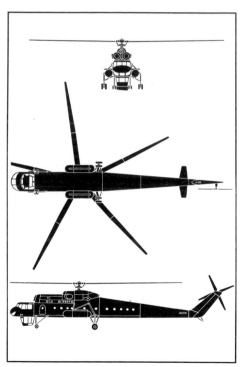

Developed from the Mi-6, the Mi-10 'Harke A' is a very large flying crane which first flew in 1960. Bulky loads can be carried between the long main undercarriage legs and the cabin accommodates up to 28 passengers. The Mi-10 is used by both the Soviet armed forces and Aeroflot. The Mi-10K 'Harke B' has the same fuselage but a short undercarriage and a gondola under the nose. A closed-circuit television system, with cameras scanning forwards from under the rear fuselage and downwards through the sling hatch, is used to monitor Mi-10 payloads. The pilot also uses this system as a touchdown reference. The Mi-10K was first flown in 1966. Range of the Mi-10 is 250km and cruising speed 180km/h. *Country of origin:* USSR. *Silhouette:* Mi-10K. *Picture:* Mi-10.

Power: 2 × JTFD 12 turboshafts   *Rotor dia:* 21.9m   *Length:* 21.41m

A flying crane designed to lift heavy bulk loads, the S-64 first flew in 1962 and was adopted by the US Army as the CH-54A Tarhe. Special standardised containers can be slung under the fuselage boom. With a maximum loaded weight of 9,072kg, these pods can accommodate 45 combat-equipped troops, 24 stretchers, cargo or other equipment. They can also be adapted in the field to act as surgical units or command and communications posts. A variant with more power and modified rotor system is known as the CH-54B. Civil versions are designated S-64E and S-64F. Range is 370km and maximum cruising speed 204km/h. Production has ceased but S-64s remain in US Army Reserve service. *Country of origin:* USA. *Silhouette and picture:* S-64.

Kamov Ka-25 'Hormone'

Confusion: 'Hoodlum', 'Helix'

Power: 2 × Glushenkov turboshafts   *Rotor dia:* 15.74m   *Length:* 9.75m

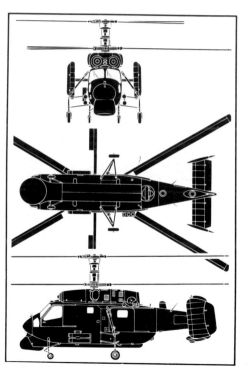

Produced in large numbers, the Ka-25 'Hormone' is the Soviet Navy's standard shipborne anti-submarine helicopter. Distinguished by a large chin radome, it carries depth charges, torpedoes and guided missiles. Twin main rotors are mounted on a single shaft. A transport version seats 12, while 'Hormone B' carries special electronics equipment and 'Hormone C' is a utility search and rescue model. The Ka-25K, with the radome replaced by a glazed gondola, is a commercial version for crane and general duties. Cruising speed is 193km/h and range 402km. *Country of origin:* USSR. *Silhouette and picture:* 'Hormone A'.

Power: 2 × Isotov turboshafts   *Rotor dia:* 15.9m   *Length:* 11.3m

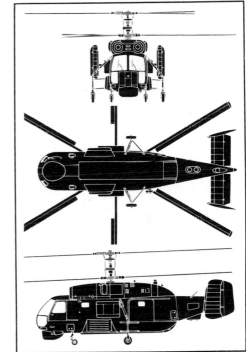

First observed aboard a Soviet cruiser in 1981, this latest Kamov helicopter is designated Ka-27 and is NATO code-named 'Helix'. The layout of 'Helix' is similar to that of 'Hormone' but the former is significantly larger and heavier, with a longer fuselage, increased rotor diameter and two instead of three fins. 'Helix' is basically an anti-submarine helicopter, with two pilots and three sensor operators. 'Helix A' is the ASW version, Ka-29 'Helix B' is the naval infantry assault version and Ka-28, the export version. A search and rescue variant ('Helix D') has been observed and the civil versions are known as Ka-32S and T. *Country of origin:* USSR. *Silhouette and picture:* 'Helix A'.

Kamov Ka-26 'Hoodlum'

Confusion: 'Hormone', 'Huskie'

Power: 2 × Vedeneev piston engines   *Rotor dia:* 13m   *Length:* 7.75m

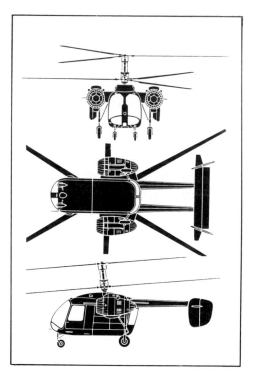

First flown in 1956, the Ka-26, NATO code-named 'Hoodlum A', is a light transport and agricultural helicopter in large-scale service in Eastern Europe and exported to such countries as West Germany, Sweden and Sri Lanka. Payloads are interchangeable by the use of specialised fuselage pods designed for the transport, ambulance, cropdusting/spraying and geophysical survey roles. Carrying seven passengers, 'Hoodlum' has a range of 400km. A single turboshaft-engined version is the Ka-126 'Hoodlum B', to which conversions are being carried out in Romania. *Country of origin:* USSR. *Silhouette:* Ka-26. *Picture:* Ka-126.

Power: 2 × T55 turboshafts   *Rotor dia.* 18.29m   *Length:* 15.54m

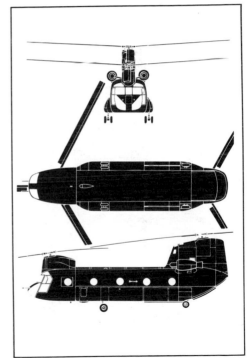

Many hundreds of tandem-rotor Chinook helicopters have been built. In addition to US production, the type is made under licence in Italy. Initial variant was the CH-47A, followed by the uprated CH-47B. The current production model is the CH-47C with more range and better performance. The US Army is upgrading all its Chinooks to a new -47D standard. The RAF operates the Chinook designated HC Mk1 and there have been exports to other countries. Up to 44 troops can be carried in the cabin and large loads can be slung. The international military version is the Model 414. The commercial passenger version, the Model 234, is in service with Columbia Helicopters in the USA. A version for US Special Operations Forces is designated MH-47E. *Country of origin:* USA. *Silhouette and picture:* CH-47C.

Boeing-Vertol H-46 Sea Knight/Model 107 *Confusion:* Chinook

Power: 2 × T58 turboshafts   *Rotor dia:* 15.54m   *Length:* 13.66m

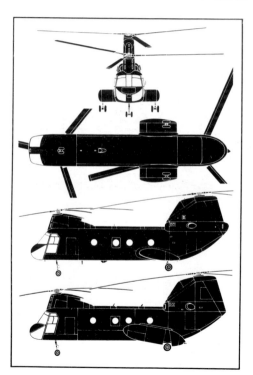

The Sea Knight operates as a shipboard and land-based assault transport carrying 25 troops or freight. First flown in 1958, the Sea Knight has been built in a variety of versions: the CH-46A, -46D, -46E and -46F for the US Marines, and the UH-46A and -46D for the US Navy. Canada operates the type under the designations CH-113 and CH-113A, while Sweden bought some Rolls-Royce Gnome-powered HKP-4s. The commercial version is the Model 107, now solely built by Kawasaki, which also supplies a military KV-107 to the Japanese forces. Range is 1,022km and cruising speed 248km/h. *Country of origin:* USA/Japan. *Main silhouette:* H-46; *lower side view:* Kawasaki KV-107. *Picture:* KV-107/II-5.

A selection of aircraft coming into service in the future and those still around but few in number

An advanced Soviet 214 seat medium range airliner with twin turbofans and fly-by-wire controls, the Tupolev Tu-204 first flew in January 1989. It is to be built in large numbers, to replace the Tu-134 and Tu-154 in Aeroflot. With a 13 tonne payload the 204 has a range up to 4600km.

A regional airliner for the 1990s the CF34 turbofan powered Canadair RJ can carry 50 passengers on routes up to 2750 km long. The company expects that up to 1000 aircraft of this class will be required to meet market demands. The RJ is derived from the Challenger executive aircraft airframe and at the time of launch in March 1989, 56 orders had been placed. (Artist's impression)

A combined four-engine/twin-engine programme the A340/A330 projects represent a logical extension of the Airbus range. The A340-300 with four engines and seating for up to 440 passengers is scheduled to fly in May 1991 followed by the longer range A340-200 in October 1991. The twin-engined 335 passenger A330 is scheduled to fly in June 1992. A combined passenger-freight variant, the A340-300 Combi is also on offer. The A340 has four CFM56 engines, while the A330 is offered with CF6, RB211 or PW4000 turbofans. Upper artist's impression A340, lower impression A330.

The Aero Industry Development Centre in Taiwan has built the prototype Indigenous Defence Fighter (IDF), a twin-engine supersonic aircraft which it is intended to build in quantity. (Upper picture)

Scheduled to fly in 1990, the McDonnell Douglas C-17A is a heavy lift transport capable of carrying up to 71,895 kg and operating out of short rough fields. The USAF plans to procure up to 210 C-17s by the end of the century. (Lower artist's impression)

Three new combat aircraft are being developed in Western Europe/Scandinavia but will not be in service in the lifespan of this edition of *Jane's World Aircraft Recognition Handbook*. All three aircraft have foreplanes and cropped delta wings. The Dassault-Breguet Rafale (prototype, above left) has two reheated turbofans, the multi-national European fighter aircraft (EFA) (artists impression above right) is also twin-engined, while the Saab JAS 39 Gripen (prototype, left) is smaller and has a single engine.

Having served with 25 nations, the F-86 Sabre, which won its spurs against the MiG-15 in Korea remains in front-line service only in South America. Many have been converted into remotely piloted vehicles. (Upper picture)

Three VFW 614s remain in service with the German Air Force for communications and VIP work. The type is unusual in having the engines mounted on pylons above the wings. (Lower picture)

A handful of Proteus turboprop Britannias remains in service, together with a few of the Canadian built CL-44s with Rolls Royce Tyne engines.

Originally flown by British European Airways, the Vanguard Tyne-powered airliner remains in service in small numbers. The freighter version is known as the Merchantman.

Originally in RAF service, the Argosy rear loading freighter is still used in small numbers for civil operations.

Heavy Lift Cargo Airlines continue to use a small number of Shorts Belfast heavy freighters on charter services round the world.

GLOSSARY OF RECOGNITION TERMS

aerofoil section Cross-section through an aircraft wing or helicopter rotor blade.

afterburner Gives a jet engine greatly increased thrust by burning fuel in the exhaust gas stream. Also called reheat.

aileron Wing-mounted lateral control surface.

airbrake Retractable 'door' mounted on the fuselage or wing and used to slow the aircraft down or steepen descent angle.

AI Airborne intercept radar.

all-up weight Fully loaded weight of an aircraft or helicopter.

amphibian Aircraft capable of alighting on both land and water.

anhedral Angle at which a wing slopes down from the horizontal in the head-on view. Opposite of dihedral.

antenna Any form of aerial for the transmission/reception of radio/radar signals.

arrester hook Underfuselage hook designed to engage wires on an airfield or a carrier deck in order to stop the aircraft.

aspect ratio The ratio between the span and chord of a wing. Derived by dividing span by chord.

ASV Air-to-surface-vessel radar. Used by maritime aircraft.

balance tab Part of an aileron, elevator or rudder which moves into the opposite sense to the main control surface.

bank/banking Manoeuvre in which the aircraft is rolled to left or right.

bicycle undercarriage Two main wheels in tandem under the fuselage. Usually with outrigger wheels on the wings.

bogie undercarriage Four or more wheels on each undercarriage leg.

camber Curvature of an aerofoil section.

canard Aircraft with its tailplane mounted forward of the wing.

canopy Transparent cockpit cover.

cantilever Wing or undercarriage without bracing struts.

c of g Centre of gravity.

chord Measurement from the leading edge to the trailing edge of a wing or rotor blade, parallel with the fuselage centreline.

contrail White vapour trail caused by the condensation of the water in engine exhaust gases at certain altitudes.

contra-rotating propellers Two propellers mounted on one shaft and rotating in opposite directions.

crescent wing Sweptback wing with varying degrees of sweep on the leading edge.

delta Triangular wing or tailplane.

dihedral Angle at which a wing slopes up from the horizontal in the head-on view. Some wings have dihedral on the outer sections only. Opposite of anhedral.

dogtooth Notch in a wing leading edge, visible in the plan view. Also known as a sawtooth.

dorsal fin Extension of the main fin along the top of the fuselage.

drag The resistance offered by the air to an aircraft or part of an aircraft passing through it.

drone Pilotless aircraft used for reconnaissance, electronic warfare or as a target. Also known as a remotely piloted vehicle (RPV).

droop Means of increasing the camber and therefore the lifting capacity of a wing by mechanically extending and lowering the leading edge.

ECM Electronic countermeasures. Means of jamming or distorting radio or radar signals.

ECCM Electronic counter-countermeasures. Means of overcoming ECM.

ejection seat Seat designed to fire aircrew clear of the aircraft in an emergency.

elevator Tailplane-mounted surface used to control an aircraft in the vertical plane.

elevon Combined elevator and aileron; used on delta-wing aircraft.

fence Vertical metal strips mounted chordwise on a wing upper surface to contain spanwise airflow.

fenestron Multi-blade tail rotor mounted in a duct.

fin Fixed part of a vertical tail.

flap Surface mounted at the wing trailing edge and deployed to increase lift for take-off and landing.

flaperon Trailing-edge surface combining the functions of flap and aileron.

flat four Horizontally opposed four-cylinder piston engine. Also flat twin and flat six.

flight refuelling Means of passing fuel from one aircraft to another in the air. Under the British system the 'tanker' aircraft trails a hose with a drogue at the end, into which a probe on the receiving aircraft is inserted. The American system is based on a 'flying boom' which can be directed to a receiving point on the aircraft which is being refuelled.

flying tail Tailplane which has no fixed portion and moves in its entirety to give vertical control.

foreplane Forward control surface on a canard aircraft.

fuselage Main body of an aircraft or helicopter.

gull wing A wing with dihedral on the inner sections only. An inverted gull wing has an hedral on the inner sections.

hardpoint Reinforced point, usually on the wings, to which external stores can be attached.

JATO Jet-assisted take-off. Technique under which auxiliary rocket packs are mounted on the aircraft to shorten the take-off run. Also known as RATO (rocket-assisted take-off).

leading edge The front edge of a wing, tail or rotor blade.

lift The upward force which keeps an aircraft flying.

Mach number The ratio between the speed of an aircraft and the speed of sound at a particular height. At sea level the speed of sound is 720 mph, so that an aircraft travelling at 720 mph would be flying at Mach 1; 576 mph under the same conditions would be equivalent to Mach 0.8. The speed of sound decreases with altitude up to 36,600ft; thereafter it stays constant at 660.6 mph.

mass balance Weight mounted internally or externally on a control surface to prevent damaging oscillation.

monocoque Aircraft structure in which the outer skin takes most of the load.

nacelle Streamlined structure housing engines, radar or other equipment.

parasol Monoplane with wings mounted on struts above the fuselage.

payload Paying load of a transport, i.e. passengers or freight.

pitot tube Instrument which detects ram air pressure. This is compared with static pressure to find airspeed.

pod Separate nacelle attached to fuselage or wing and carrying an engine, radar or other equipment.

pressurisation The maintenance of pressure inside an aircraft so that passengers and crew can breathe without personal oxygen supplies at high altitude.

pylon Wing or fuselage-mounted structure carrying engine pods, auxiliary fuel tanks or weapons.

radome Dome-shaped fairing over a radar aerial.

refuelling probe Used to receive fuel from an airborne tanker.

reheat See *Afterburner*.

rotor blade Helicopter lifting and control surfaces. A rotor is made up of two or more blades. The main rotor generates lift and creates propulsive thrust. The vertically mounted tail rotor balances out the torque which makes the fuselage tend to rotate in the opposite direction to the main rotor.

rudder A fin-mounted vertical surface which provides directional control.

servo tab Moving surface on aileron, elevator or rudder which when moved causes the main control surface to deflect in the opposite direction.

slat Section of wing leading edge which moves forward to produce a slot between itself and the wing. Increases lift and improves control at low speeds.

slot Spanwise gap at the wing leading edge (see *slat*).

slotted flap Flap which creates a slot between itself and the wing trailing edge.

sonobuoy Listening device dropped from an aircraft to detect submarines.

split flap Flap which is hinged forward of the wing trailing edge, so that the wing upper surface is unchanged when the flap is lowered.

spoiler Retractable wing-mounted surface deployed to create drag and reduce lift. Used to slow the aircraft and control its descent path.

sponson Stub wing carrying undercarriage or other equipment.

spring tab Small control surface which acts as both balance tab and servo tab.

stabiliser Tailplane (US).

STOL Short take-off and landing.

subsonic Speeds below that of sound, i.e. less than Mach 1.

supercharger Compressor used to boost the power of a piston engine. Also called a turbocharger.

supersonic Speeds in excess of sound, i.e. above Mach 1.

sweepback Plan-view angle between the wing and the fuselage centreline.

taileron Tailplane used as a primary control surface in both pitch and roll.

tailplane Horizontal surface at the rear of an aircraft which normally carries the elevators.

taper The decrease in wing chord from roof to tip.

thickness/chord ratio Ratio of the thickness of the wing section to the chord at a given point. Usually expressed as a percentage.

thrust The force that moves an aircraft forward. Also used to express the power of a jet engine.

thrust reverser System which turns the thrust of a jet engine forwards for braking purposes.

transonic Speeds just above and below Mach 1.

trim tab Adjustable tab on a control surface.

turbofan A jet engine with a large fan on the front which passes air round the core of the engine and also gives more impetus to the compressor.

turbojet An engine in which the hot gases expelled at the rear provide all the propulsive thrust.

turboprop A gas turbine engine in which the shaft is geared to drive a propeller.

turboshaft A gas turbine in which the power is transmitted to a shaft, usually to drive a helicopter rotor.

vortex generators Small protruding plates fitted to wing, tail or fuselage to change the airflow pattern.

wing loading Aircraft loaded weight divided by wing area.

Test your aircraft-recognition skill by identifying these sillograph outlines of well known aircraft. Tackle the whole test in one go, and write down your answers. The solution is on the last page.

SILLOGRAPH PROFICIENCY TEST

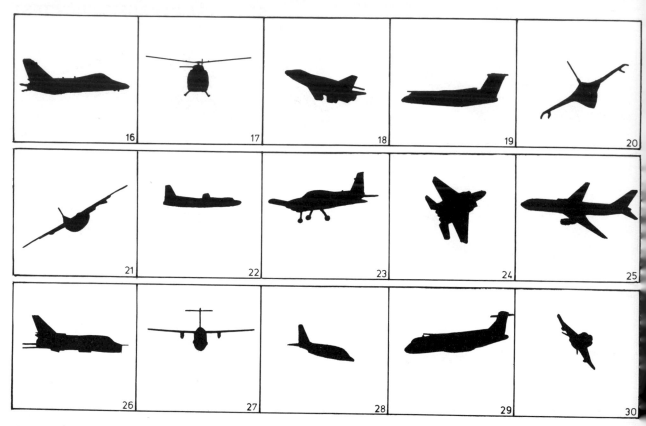

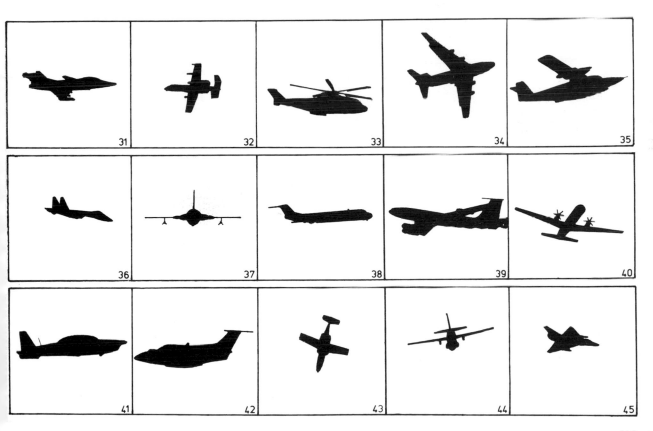

Test answers (see page 583)

1 Mi-24 'Hind'
2 Mirage F1
3 Su-25 'Frogfoot'
4 Sea Harrier
5 AH-1 HueyCobra
6 MiG-29 'Fulcrum'
7 B-1
8 Tu-22 'Blinder'
9 Il-76 'Mainstay'
10 Tornado
11 Tucano
12 MiG-25 'Foxbat'
13 Dornier 228
14 Hawk 200
15 A-4 Skyhawk
16 AM-X
17 Lynx
18 Su-24 'Fencer'
19 Dash 8
20 F-16
21 An-12 'Cub'
22 F-27 Friendship
23 Valmet L-70 Vinka

24 F-15 Eagle
25 Boeing 767
26 Su-22 'Fitter K'
27 ATR 42
28 Alpha Jet
29 ATR 42
30 Phantom
31 JAS-69 Gripen
32 A-10
33 EH-101
34 An-24 'Condor'
35 Dornier Seastar
36 Su-27 'Flanker'
37 Saab Draken
38 Fokker 100
39 KC-135 Stratotanker
40 BAe ATP
41 Epsilon
42 EMBRAER Xingu
43 Saab 105
44 Transall
45 Mirage III